GEOMETRY EXAMPLES

POSITIONS AND ANGLES

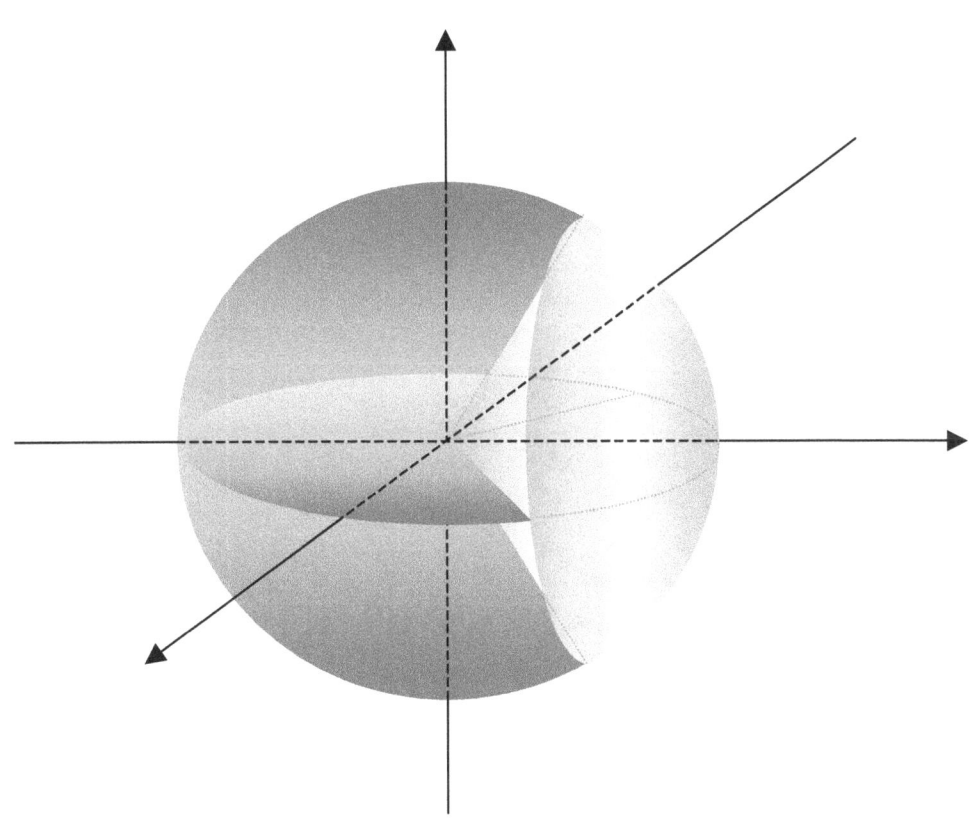

Seong R. KIM

Dear students:

Students need the best teacher, so you need examples, because examples are the best teacher. All the examples here are fully worked, and explain **how** the basic and essential tools in math are made, together with **what** they are, **how** they work, and **how** to work with them. Such tools include numbers, formulas, identities, equations, laws, etc.

Examples here begin with easy ones, of course. Covering every meter and yard properly, we can cover thousands of miles and kilometers. And it is particularly the case in math.

Of those examples therefore, some might even look too easy for you. It's not that easy though, to come up with those examples. Anyways, the bigger and the taller the tree, the deeper and the stronger the root.

Doing math, we work with ideas and run ideas, because every thing in math is an idea. A number is an idea, for instance, and the same is true for a line or circle, too. And putting ideas together, we build another, which becomes the base or an element of another, and each is connected. And that's the way your math grows. So you get to build a circuit, and sometimes, need to fill the gap or repair the circuit so that you get the sense of it.

So your calculation runs properly, and you get the problem solved.

The examples have been made and arranged so that they get tougher (or sometimes easier for some reason) as you proceed with them. In particular, similar examples with some variations are strategically repeated so that you can get the ideas or the tools tricky or complicated, and can get them mastered.

This book is however, nothing but a bunch of examples until you get it powered.   How then, to get it powered, and make it run and work for you?

Just read it, and then, do each example in writing. And it is important to note that you do it in **your** writing. Just watching someone doing it, you just only feel that you can do it. If you do it, you can do it, but if you don't, we can hardly. It's a cliché, of course, but is always true that knowing is one thing and doing is another.

I've been helping students grow, take care of, and run their own math. The area covers algebra and geometry for high school or college students, and is especially for equations (for unknowns or curves), functions, and their graphs, which are the basic elements in calculus, which's been the core of my interest from my early age in high school.

Of my students, some are quite poor in math, and thus, are afraid of or hate math, some require special education because of exceptional intelligence, some are smart enough, some are naïve and diligent, some are clever but lazy, and most behave in general. All the students are badly after though, one thing in common: a strong and secure math skill. It is of course, the prime objective of my work, and I'm always happy to and eager to help them achieve it. The problem was however, that many of them wanted it to be purchased. And the question is, can we buy it?

We can buy the means, of course. And a solid math skill is feasible, too. We know however, we can't buy love, and the same is true for the math skill, too. It's not what we can buy or sell, and not what we can give or take. It is however, what we can grow, and need to grow. Your math grows as much as you grow and take care of it. So does mine.

What math then, do students most often do or use in high schools or colleges?

It is algebra and geometry. What algebra though?

Elementary algebra, of course
Doing the algebra, we work with numbers (many in kinds), constants, variables, ratios, rates, expressions, equations, inequalities, functions, identities, formulas, laws, etc., together with signs and symbols. And if we want to do algebra properly, we want to know their natures and how they mingle with each other.

So studying math ideas or tools, you want to know **what** they are, **how** they work, and **how** to work with them or **what** to do with them.   What then, about the geometry?

Basically, the geometry has much to do with shapes, positions, and angles. The shapes begin with triangles and circles, and move on to rectangles, squares, parallelograms or rhombuses, trapezoids, tetragons, other polygons, polyhedrons, etc.

Doing the geometry, too, though, we need to do the algebra stated above. So it is analytic geometry, often called coordinate geometry, too. And doing it, we can specify positions using coordinates. So in the geometry, basically, we work with graphs. Putting a math idea in a graph, we can not only effectively think about it but actually see it, too, and therefore, can efficiently work with it. What idea then, is it?

The idea begins with a point, line, parabola, circle, ellipse, and hyperbola, called a conic section or basic curve, and then, moves on to other curves, planes, surfaces, volumes, and other objects in various dimensional spaces, together with vectors.

And using an angle, we can specify an amount of turn or change in direction.

So learning, using, or applying those ideas or math tools, we get to solve problems.

And this book can help. It can help learn them, and use them so that you can navigate to find solutions to problems. And in particular, it can help come up with answers to those **what**s and **how**s stated above. So it can help you grow and run your own math, and thus, can help achieve your solid math skill.

It is however, not a magic book giving you a math skill of high caliber overnight. And it can have many mistakes, too. There is no magic, and math is full of facts and ideas. And it is after all, not me and not your teacher but you who put together some of those facts and ideas, and understand it. Putting facts and ideas together, understanding it, and taking care of what you have learned, you grow your math. And this book can help.

This is a book of examples designed to help you grow your math, and assumes that you are a real beginner. This book requires though, time and effort, the amount of which need to be substantial, too, but will be worth it. That's because you want a substantial achievement, and will get it. And probably, you will get to see this book helping you get there much faster than expected. And then, you will get to see the way math runs.

In math, everything is an idea. So is a problem. And solving it, we put it many different ways. For instance, while expanding or reducing it, or modifying or converting it, we keep searching for the solution, approaching the solution, and eventually, can get there. So don't look for the solution outside the problem. The solution is inside the problem if the problem is properly made.

If it is not, no solution is the solution. And in fact, it is often the case a problem itself is the solution. We can put a problem in many different ways, and eventually, can end up with the solution.   How come then, is the solution no other than the problem?

For instance, the solution to $3232 \div 101$ is 32.   And we can put it this way:

$$3232 \div 101 = \frac{3232}{101} = \frac{32 \times 101}{101} = \frac{32}{1} = 32 \implies 3232 \div 101 = 32.$$

And we can get this, too: $32 \implies 3232 \div 101$.   How?

$$32 = \frac{32}{1} = \frac{32 \times 101}{101} = \frac{3232}{101} = 3232/101 = 3232 \div 101.   \text{Too easy?}$$

For another instance, the solution to $ax^2 + bx + c = 0$ is: $x = \frac{-b \pm \sqrt{b^2 - 4ac}}{2a}$, which is called the quadratic formula.   How come then, is the solution no other than the problem?

We can put it this way:

$$x = \frac{-b \pm \sqrt{b^2-4ac}}{2a} \implies 2ax = -b \pm \sqrt{b^2 - 4ac} \implies 2ax + b = \pm\sqrt{b^2 - 4ac}$$

$$\implies (2ax + b)^2 = b^2 - 4ac \implies 4a^2x^2 + 4abx + b^2 = b^2 - 4ac$$

$$\implies 4a^2x^2 + 4abx = -4ac \implies ax^2 + bx = -c \implies ax^2 + bx + c = 0.$$

And we can get this, too: $ax^2 + bx + c = 0 \implies x = \frac{-b \pm \sqrt{b^2-4ac}}{2a}$.   How?

$$ax^2 + bx + c = a(x^2 + \tfrac{b}{a}x) + c = a(x^2 + \tfrac{b}{a}x + \tfrac{b^2}{4a^2} - \tfrac{b^2}{4a^2}) + c = a(x^2 + \tfrac{b}{a}x + \tfrac{b^2}{4a^2}) - \tfrac{b^2}{4a} + c$$

$$= a(x + \tfrac{b}{2a})^2 - \tfrac{b^2-4ac}{4a} = 0 \implies a(x + \tfrac{b}{2a})^2 = \tfrac{b^2-4ac}{4a} \implies (x + \tfrac{b}{2a})^2 = \tfrac{b^2-4ac}{4a^2} \implies x + \tfrac{b}{2a} = \pm\sqrt{\tfrac{b^2-4ac}{4a^2}}$$

$$\implies x = -\tfrac{b}{2a} \pm \tfrac{\sqrt{b^2-4ac}}{2a} = \tfrac{-b \pm \sqrt{b^2-4ac}}{2a} \implies x = \tfrac{-b \pm \sqrt{b^2-4ac}}{2a}.$$

And we call the set of processes above, algebra.

So if a problem is well defined, that is, if it makes sense, we should be able to get it solved the way below:

A problem $\Rightarrow$ **...** $\Rightarrow$ **...** $\Rightarrow$ **the solution**, and thus: **the problem** $\Rightarrow$ **the solution**.

So solving a problem, we put it many different ways so that we can get to the solution.

And that's the way, math runs.

May your math run very well.

Seong R. Kim

B.S. Math. Michigan Tech. Univ.   M.S. Math. Rensselaer Polytechnic Institute

Notes:

This book is a book of examples, and the examples are about basics in geometry. In particular, this book is about angles, and explains how to work with angles and how to find angles doing geometry.   Why angles though?

Doing problems in geometry, we usually get to find lengths or distances, areas, or volumes, and angles, too. Finding an area of a triangle, for instance, we can use the formula where the area is half the product of the base and the height. What if however, the height is not given?

If knowing the angle between the base and a side in the triangle, you can get the height. You can get it using trigonometry.

Knowing the angle between the one you want to find and an object known to you, you can find the one you want, and can get it using trigonometry.

Doing geometry, we often need to know how things are positioned, which means we want to know what angles they have, even though the their distances are given.

What is an angle though?   What is it about?   What do we do with it?   And how?

If not knowing about angles, you want to know what angles are, what they are about, and what to do with them so that you can find it when you need it. And that is what this book is about.

And thus, this book explains what angles are, how they work, and what to do with them, together with how to do it. So you are going to learn those so that your geometry can run not only properly but fast enough, too.

Contents

Note:

The drawings or graphs in this book are not exact, and are approximate or conceptual ones.

$\in$	"$a \in B$" means that a belongs to B. "$p, q,$ and $r \in W$" means that $p, q,$ and r belong to W.
$\Rightarrow$	"$A \Rightarrow B$." means that A implies B.
$\equiv$	$A \equiv B$ means that A and B are identical to each other.
$\neq$	$A \neq B$ means that A is not equal to B.
$\lvert A \rvert$	The magnitude of A. For instance, $\lvert -1 \rvert = \lvert 1 \rvert = 1$.
$\therefore$	Therefore
$\Leftrightarrow$	"$A \Leftrightarrow B$" means "If A then B." and "If B then A." We can read $A \Leftrightarrow B$ as "A if and only if B." In such a case, we can say that $A = B$.
Δx and Δy	Suppose that (x_1, y_1) and (x_2, y_2) are two points in the x-y plane. Then, we get either of the two below. $\Delta x = x_2 - x_1$, and $\Delta y = y_2 - y_1$. $\Delta x = x_1 - x_2$, and $\Delta y = y_1 - y_2$.

Distance Formula

Suppose that d is the distance between two points (x_1, y_1) and (x_2, y_2) in the x-y plane. Then, we get: $d^2 = (\Delta x)^2 + (\Delta y)^2$.

0. **What is a triangle?**

To begin with, tri means three, so literally, a triangle is made of three angles.

Fig. 0.0

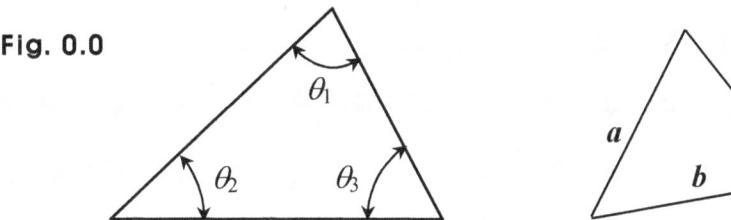

And next, a triangle is made of three sides, and is a polygon. What is a polygon though?

It is a 2-D object closed, and is made of three or more line segments. Usually though, we just call each line segment a side. And in basic math, we normally work with polygons said to be simple and convex. What then, is such a polygon?

In each side in a polygon simple and convex, each endpoint is an endpoint of one of the other sides. So for instance, the polygons below are not simple and convex.

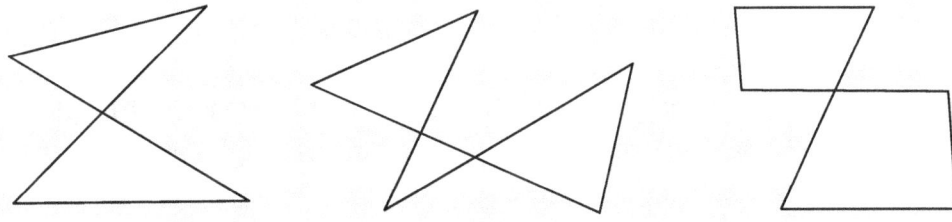

That's because in each polygon above, some side has an endpoint that belongs to more than one of the other sides.

And a line passing through a polygon simple and convex can cross up to two sides in the polygon, so it cannot cross three or more sides. How?

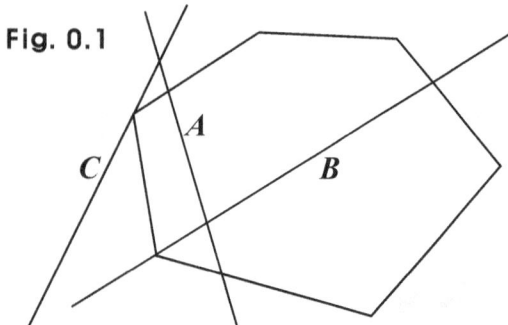

Fig. 0.1

The line *A* crosses two sides in the hexagon.
The line *B* passes through a point in the hexagon, but crosses one side only.

The line *C* meets the hexagon at a point, but crosses no side.
And no line can cross more than two sides in the hexagon.

And among such polygons, we have triangles, quadrangles as rectangles, squares, rhombuses, parallelograms, and trapezoids, and others as pentagons, hexagons, etc.

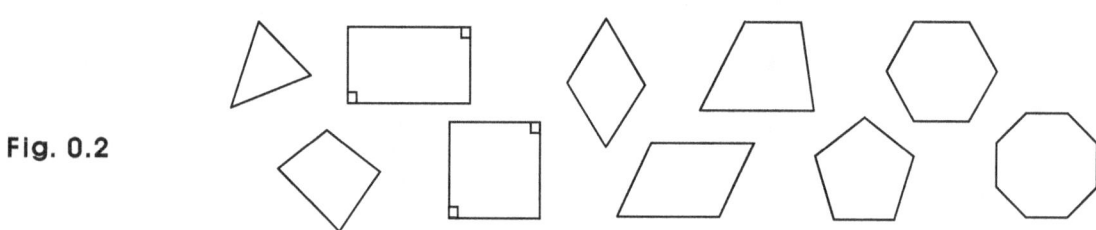

Fig. 0.2

So what can we say about triangles?

A triangle is a polygon made of three sides, in each of which, each endpoint is an endpoint of one of the other two sides. And thus, triangles are the simplest polygons.

No matter what polygon it may be though, it can be said to be made of triangles. More specifically, it can be partitioned into triangles. And thus, triangles are not only the simplest but the most basic polygons, too.

And we can say also, a triangle is made of many other triangles, too, in each of which, one angle is 90°, called a right angle.
So those other triangles are called right triangles.

More specifically, no matter what triangle it may be, it can be partitioned into two right triangles at a time.
And we can partition a right triangle into two other right triangles at a time, also.

And thus, it can be said that every polygon is made of right triangles.

Fig.0.3

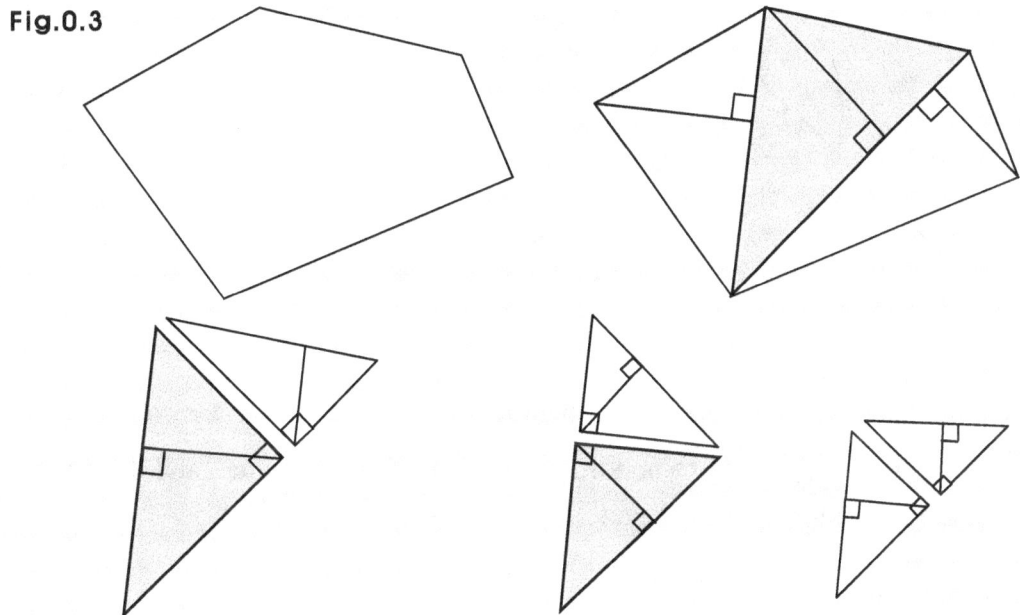

So what can we say about right triangles?

Right triangles are fundamental triangles, and thus, are important. And they are very much so. It is often the case we can't do much without right triangles solving problems.

And we can do a lot using right triangles.
Right triangles do much not only in geometry but in algebra, too. It is often vital that we use right triangles right.
And thus, we want to know them very well, and use them very well, too.

4

Fig. 0.4

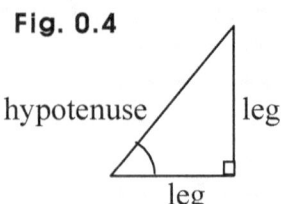

hypotenuse / leg / leg

A right triangle is made of three sides, called a hypotenuse and two legs. And the two legs are perpendicular to each other, and thus, makes 90°. So the hypotenuse is facing the angle 90°, and thus, is opposite of the right angle. And the sum of the two angles adjacent to the hypotenuse is 90°, because the sum of all the three angles in a triangle is 180°. How come it's 180° though?

To begin with, putting a triangle on a line called x, we can put it the way below:

Fig. 0.5

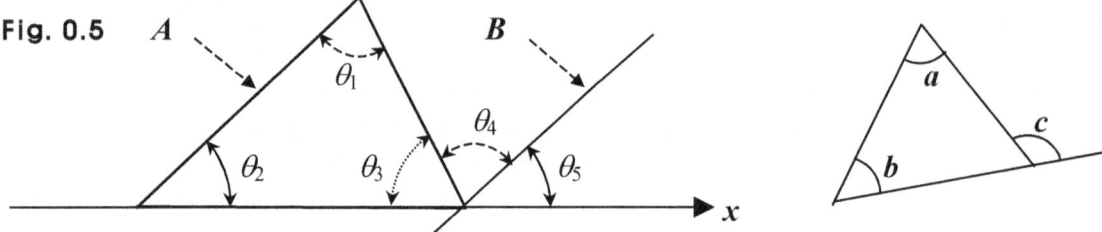

And assuming next, the line segment A is parallel to the line B above, we can say that:

The two angles θ_2 and θ_5 are corresponding angles, and thus, are equal, and the two angles θ_1 and θ_4 are alternate angles, and thus, are equal, too.

So we get: $\theta_1 + \theta_2 + \theta_3 = \theta_3 + \theta_4 + \theta_5$, which is 180°.

That is to say that the sum of all the three internal angles in a triangle is 180°.

And we can notice that in the triangle on the right in the figure above, the sum of two angles a and b is the same as the angle c. In a triangle, in fact, the sum of two internal angles is the same as the external angle supplement to the other internal angle. If two angles are supplement to each other, the sum of the two is 180°.

What then, about the sum of all the three external angles?

We know if two line segments are in a line, the angle between the two is 180°.

Fig. 0.6

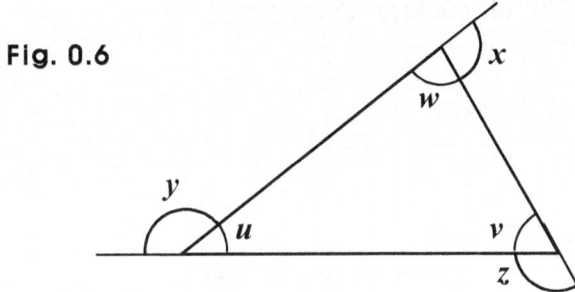

And we know: $x + y + z = 180^{\circ}$.

So the sum is: $3 \cdot 180^{\circ} - 180^{\circ} = 360^{\circ}$.

And thus, a triangle has three angles and three sides, and is a polygon with the smallest number of angles and sides. So it is a polygon the simplest and the most basic, and thus, is the most important of all polygons.

And no matter what triangle it may be, what does its three angles add up to?

The three angles add up to 180°. And we can partition any triangle into two right triangles at a time. That is to say that the sum of all the three angles in a triangle is 180°, and every polygon can be made of right triangles.

And we want to note that a triangle is an idea and not a material object, and that saying just a triangle, we mean an object made of three line segments and not a triangular disk, which is full of points, and thus, is a plate. So *nothing* inside a triangle is a part of the triangle, which is therefore, made of three line segments only, and is empty inside.

Fig. 0.7

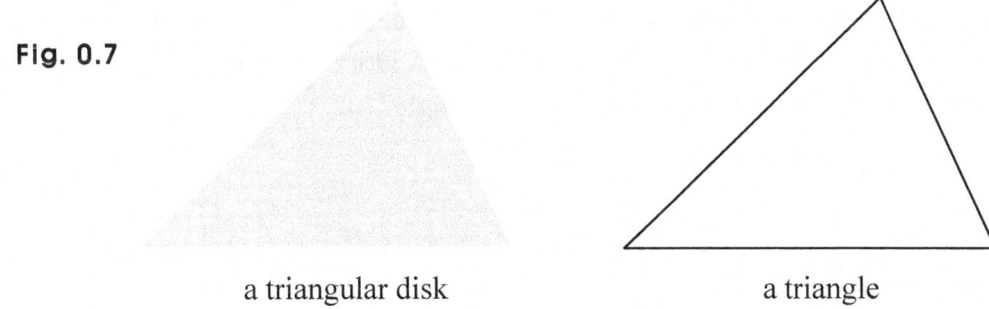

a triangular disk   a triangle

And in a triangle, if all the three angles are different, how are all the three sides?

If in a triangle, all the three angles are different, all the three sides are different, too.

And of course, if all the three sides are different, all the three angle are different, also. And we call such a triangle a *scalene* triangle.

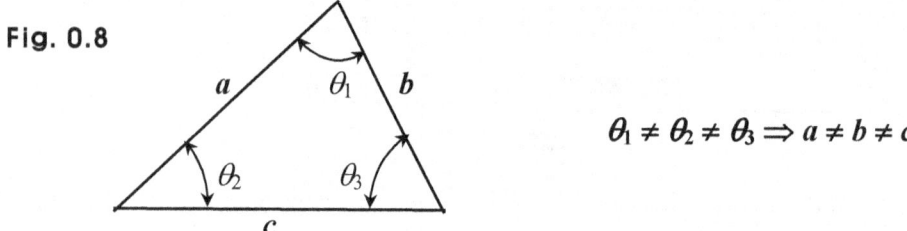

Fig. 0.8

$$\theta_1 \neq \theta_2 \neq \theta_3 \Rightarrow a \neq b \neq c$$

Next, if in a triangle, two angles are the same, its two sides are the same, too, and vice versa. And it is called *isosceles*. So an isosceles triangle is a triangle where two sides or two angles are the same. And thus, in a right triangle isosceles, what are the three angles?

In a right triangle isosceles, two angles are $45°$ each, because the other angle is $90°$, and all the three angles add up to $180°$.

Fig. 0.9

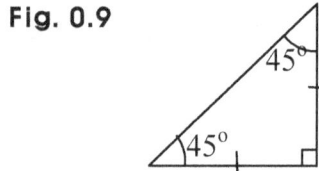

And of course, if all the thee angles are equal, all the three sides are equal, too, and vice versa. And we call it a *regular* or *equilateral* triangle. And we know the sum of the three angles is $180°$. So what is each angle in a regular triangle?

Every angle in a regular triangle is $60°$.

Fig. 0.A

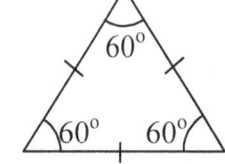

So can we form a triangle, connecting three points, using three line segments, of course?

We can form a triangle, connecting three points, only if the three points are *not* in a line. So given three points in a line, we cannot make a triangle connecting the three points.

Fig. 0.B

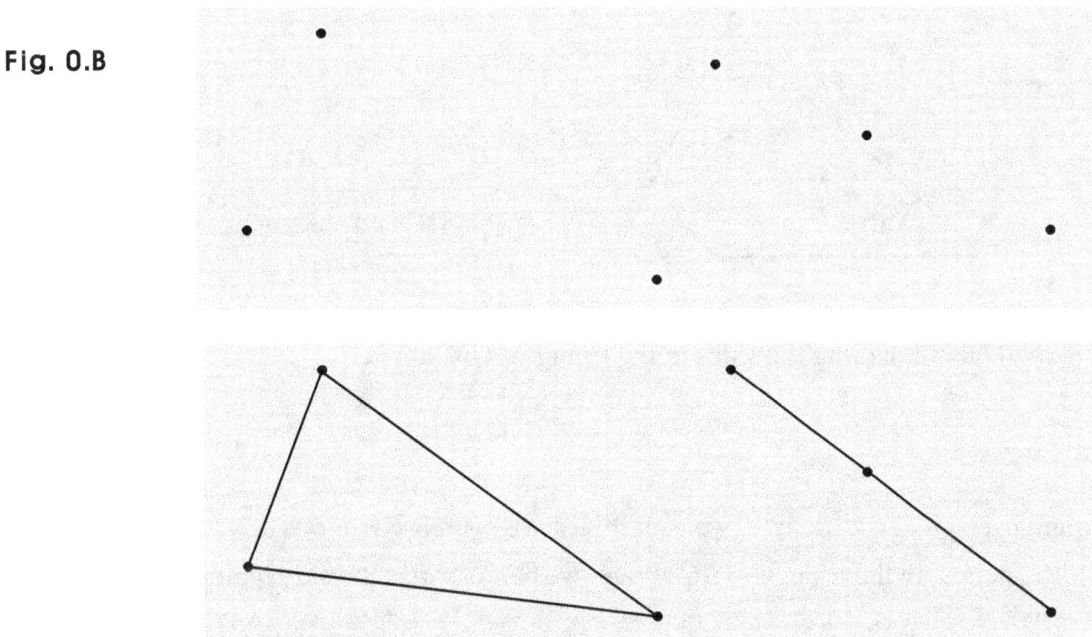

And when we work with a triangle, it's good idea to name the triangle.
And naming a triangle, we often use the names of the three vertices in the triangle.
What do we mean by the vertex though?

A vertex in a polygon is a point where two sides meet. So a triangle has three vertices.
And for instance, calling a triangle *ABC*, we mean *A*, *B*, and *C* are its three vertices.

And specifying a triangle, we often use a symbol, together with the names of the vertices.

And the symbol is Δ, which is just a small triangle.
So for instance, **Δ*ABC*** means a triangle ***ABC***. What then, about naming each angle?

Naming an angle a triangle has, we usually use a capital letter as *A*, and the angle is an internal angle, of course. That is to say that naming an angle in a triangle, we use the name of the vertex that has the angle.

So for instance, saying the angle *A* in a triangle *ABC*, we mean the angle at the vertex *A*. And indicating such an angle, we often use a symbol called an angle symbol, which is ∠. So for instance:

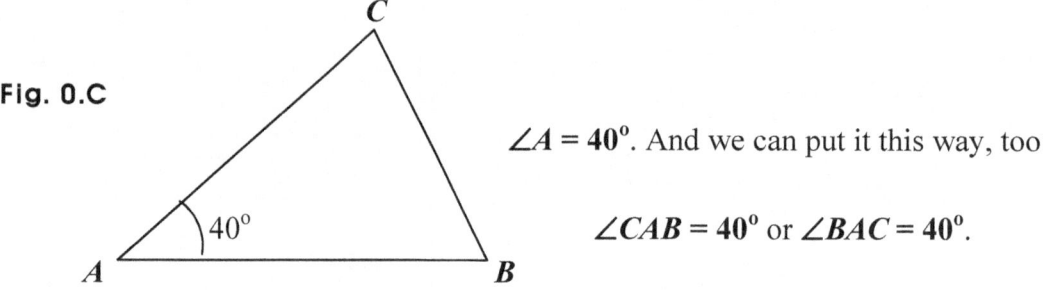

Fig. 0.C

∠*A* = **40°**. And we can put it this way, too:

∠*CAB* = **40°** or ∠*BAC* = **40°**.

What then, about naming the sides in the triangle *ABC* above?

Naming each side in a triangle, we usually use a lowercase letter as *a*.

So for instance, in the triangle *ABC* above, we can use *a* as the side facing the angle *A* or the vertex *A*. That is, *a* is the side opposite of the angle *A* or the vertex *A*. So for instance, we can put the names of the angles, the vertices, and the sides the way below:

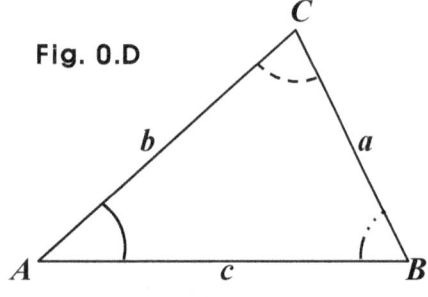

Fig. 0.D

And thus, the side *b* is the side facing the angle *B* or the vertex *B*, and *c* is the side facing the angle *C* or the vertex *C*. And of course, indicating *c*, we can indicate it by *AB*, too, which is the same as *BA*, of course. And the angle *B* can be indicated by ∠*B*, and the angle *C* can be indicated by ∠*C*, of course.

And we can put a triangle in one of three kinds: acute triangles, right triangles, and obtuse triangles. What are those three kinds though?

To begin with, in an acute triangle, all the three angles are between 0 and 90°. And such an angle is called an acute angle. For instance, a regular triangle is a triangle acute. Next, in a triangle obtuse, one angle is between 90° and 180°, and such an angle is called an obtuse angle. And as stated above, a right triangle has an angle of 90°, which is called a right angle.

Fig. 0.E

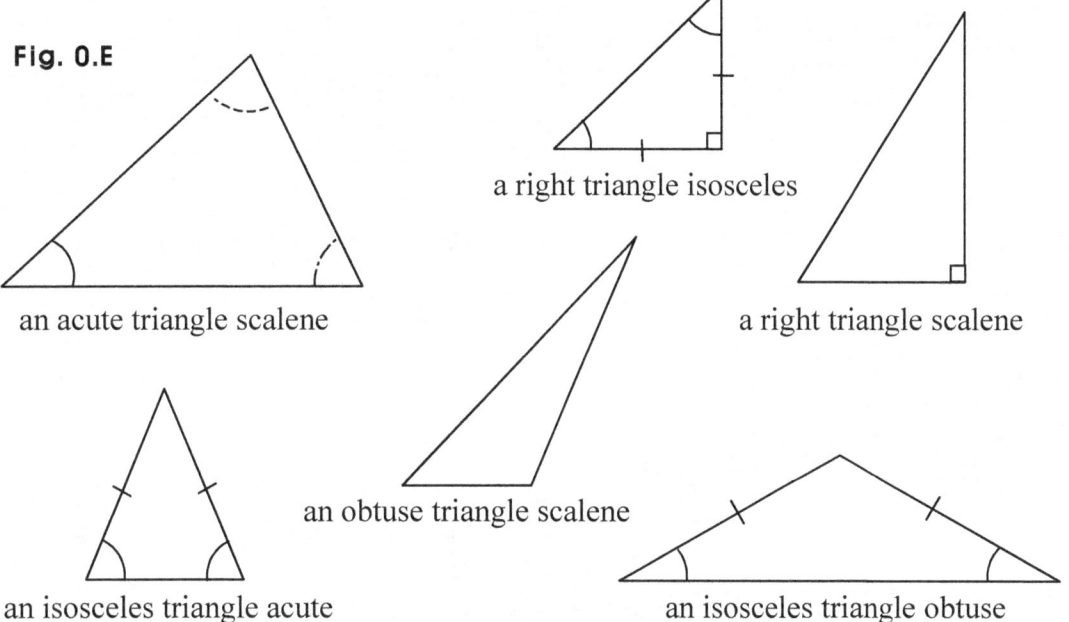

a right triangle isosceles

an acute triangle scalene

a right triangle scalene

an obtuse triangle scalene

an isosceles triangle acute

an isosceles triangle obtuse

And a regular triangle can be taken as an isosceles triangle, too, because it has two sides equal, since all the three sides in it are the same.

So we have some kinds or categories in triangles, which are the most basic and the simplest of all polygons. And putting together triangles, we can form a polygon.

That is to say that we can partition a polygon into triangles. And among those polygons, we have tetragons (quadrangles), pentagons, hexagons, etc. Tetra or quad means four, so a tetragon is a four-sided polygon. And some examples of tetragons can be as follows:

Fig. 0.F

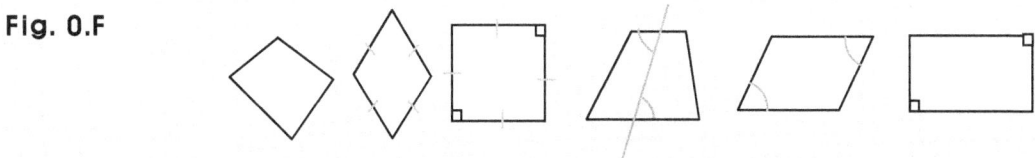

And also, putting together triangles, we can form another kind of geometric object called triangle polyhedrons. In such a polyhedron, each face is a triangle. And for instance:

Fig. 0.G

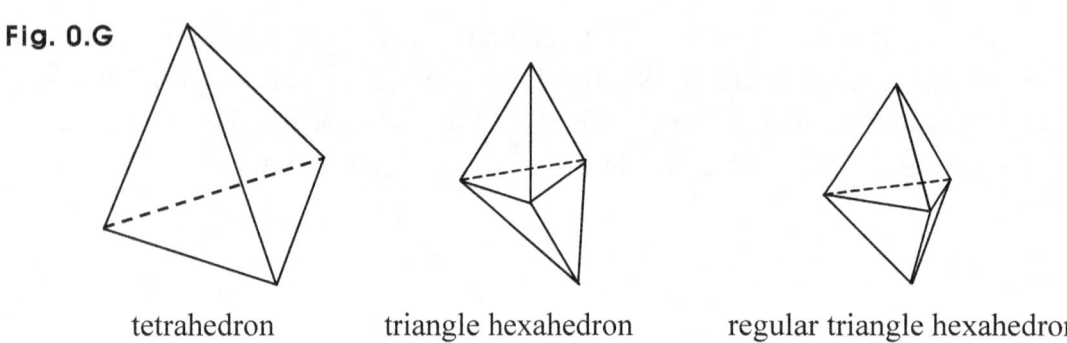

| tetrahedron | triangle hexahedron | regular triangle hexahedron |

And if each face is a quadrangle in a hexahedron, we can call it a quadrangle hexahedron.

Fig. 0.H

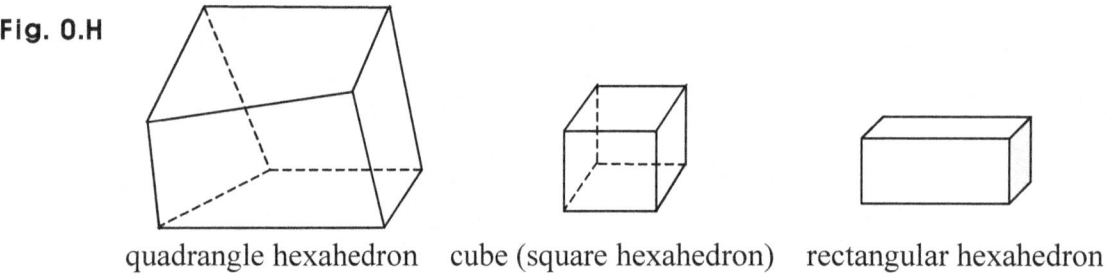

quadrangle hexahedron   cube (square hexahedron)   rectangular hexahedron

And thus, solving problems with geometry, we can work with quite a few geometric objects. And it can be said that each of those begins with a triangle.

1. Triangle Inequality

So forming a triangle, can we just use three line segments? That is, given three line segments, can we make a triangle putting together the three line segments?

It is *not always* the case where we can make a triangle connecting three line segments.

If the sum of any two of the three is larger than the other, we can make one. That is to say that if any of the three is greater than the sum of the other two, we cannot make it.

So for instance, if a, b, and c are three line segments in a triangle, we get either of the cases as follows:

• Assuming c is the biggest, we get: $a + b > c$.   For instance, $2 + 4 > 5$.

• If a is the biggest, and $b + c > a$, three line segments a, b, and c can form a triangle.

• If three line segments a, b, and c form a triangle, and b is the longest, we get: $c + a > b$.

And if we don't know which of a, b, and c is the biggest, we have to get:

$a + b > c, b + c > a$, <u>and</u> $c + a > b$.   Note that it says not or but <u>and</u>.

So if $a + b > c, b + c > a$, <u>or</u> $c + a > b$, it can be the case we get no triangle.

For instance, if $a = 2$, $b = 3$, and $c = 6$, we don't get a triangle, because we don't get this: $a + b > c$, although we get these: $b + c > a$, and $c + a > b$.

For another instance, if $a = 3$, $b = 5$, and $c = 2$, we don't get a triangle, because we don't get: $c + a > b$, although we get: $a + b > c$, and $b + c > a$.

And for another instance, if $a = 6$, $b = 3$, and $c = 3$, we don't get a triangle, because we don't get this: $b + c > a$, even though we get these two: $a + b > c$, and $c + a > b$.

Therefore, only if the three cases $a + b > c$, $b + c > a$, <u>and</u> $c + a > b$ are all true at the same time, we can get a triangle, that is, a, b, and c form a triangle.

So a triangle is not just made of three sides, and if need to check to see if three line segment can form a triangle, we have to check all the three cases above if not knowing which of the three is the biggest.

If knowing the biggest, we can see if the three can form a triangle taking the sum of the other two and comparing the sum with the one biggest. If the sum is the larger, we get a triangle; otherwise, we don't.

And the fact above is a property of a triangle, and we call it <u>Triangle Inequality</u>. And we can put the fact in a figure the way below:

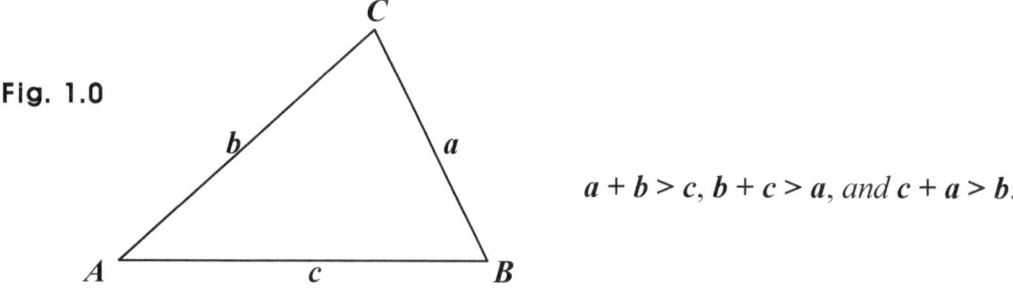

Fig. 1.0

$a + b > c$, $b + c > a$, *and* $c + a > b$.

So if a, b, and c form a triangle, we get: $a + b > c$, $b + c > a$, *and* $c + a > b$.

And using the fact above, we often find the solutions to problems not only in geometry but in other areas of math, too. So you may want to keep it in mind.

Sometimes though, it might seem the sum of two sides can be the same as the other side. For instance, suppose this time, we cut a triangle the way below.

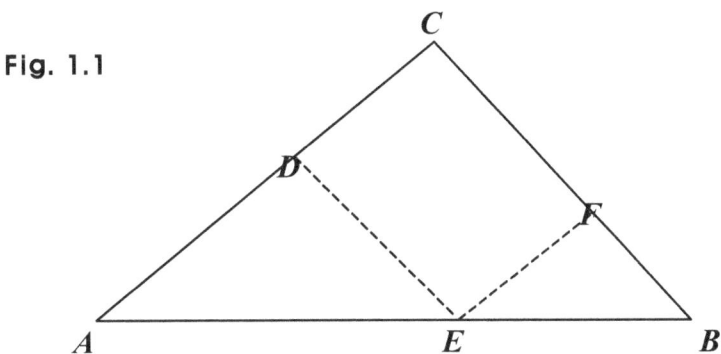

Fig. 1.1

Suppose next, *DE* is parallel to *CF*, and *EF* is parallel to *DC*.
Then, we get: $AC + CB = AD + DE + EF + FB$.

Suppose now, that we keep making more of the parallel line segments the way above.

Then, we get:

Fig. 1.2

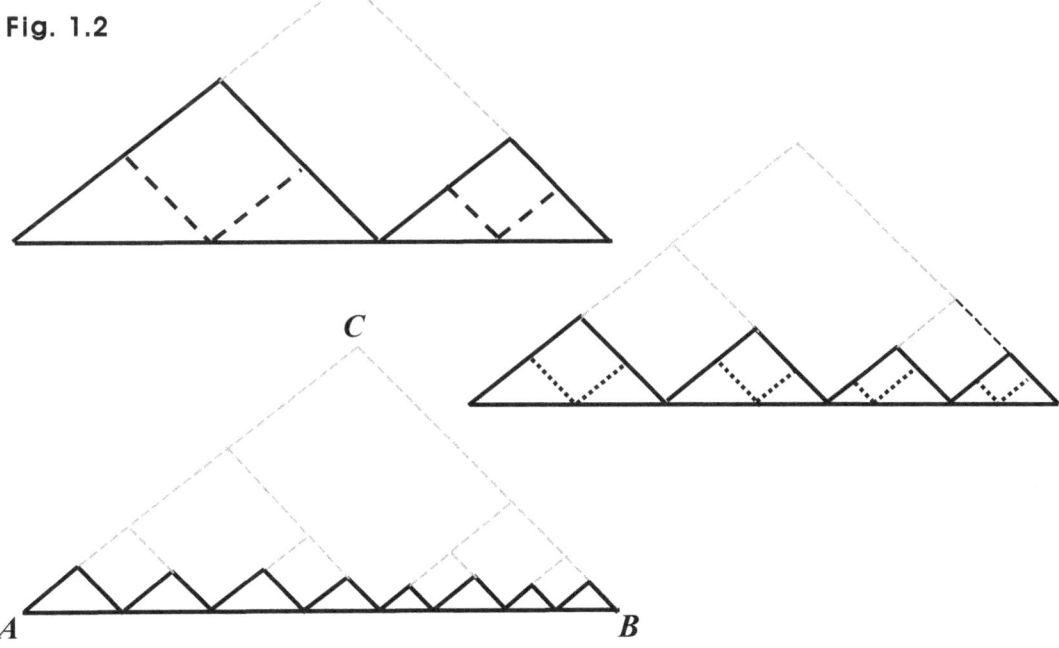

Then, it seems if we keep doing the same process as above, the sum of the two sides *AC* and *CB* is getting close to the length of *AB*. Do we then, eventually get: $AC + CB = AB$?

It is not the case, of course. That is to say that we still get: $AC + CB > AB$.

No matter how many times we may do the same processes as above, the sum of all the small line segments will be the same as the sum of the two sides AC and CB.

That is to say that the sum of all the small line segments will not converge to AB.
That's because like any other object in math, a triangle is an idea, too.
Nothing in math is a material object. Working with geometric objects as lines, triangles, circles, etc., we can get fooled by their looks.

If for instance, we make an actual triangle using wooden rods, and keep cutting two of the three and putting the pieces together the way above, there will be losses in the lengths of the two because of the waste as sawdust.

In math however, we won't get any loss in length or whatsoever no matter how many cuttings we may do to the triangle, because a triangle in math is an idea. And the same is true, too, for any other object as lines, circles, rectangles, etc.

And we have a property in the real number system, and the property is quite close to the triangle inequality, and is as follows:

- Assuming P and Q are real numbers, we get: $|P| + |Q| \geq |P + Q|$.   How come?

Assuming first, $PQ < 0$, in other words, P and Q have opposite signs as $P > 0$ and $Q < 0$, we get: $|P| + |Q| > |P + Q|$.

For instance, if $P = 2$, and $Q = -1$, we get: $|2| + |-1| = 2 + 1 = 3 > |2 + (-1)| = |2 - 1| = 1$.

Assuming next, $PQ \geq 0$, in other words, P and Q have the same sign as $P < 0$ and $Q < 0$, or $P = Q = 0$, we get: $|P| + |Q| = |P + Q|$.

For instance, if $P = -2$, and $Q = -1$, we get: $|-2| + |-1| = 3 = |(-2) + (-1)| = |2 + 1| = 3$.

And of course, we can use the fact above solving many problems, and can use it doing algebra, too. So, together with the triangle inequality, we may want to keep in mind the property above.

2. Right Triangles

Triangles are like integers. And they are very much so if they are right triangles.

We use integers to form all other kinds of numbers as 1.2, 1/3, $\sqrt{5}$, $-\sqrt[4]{3}$, etc. And the same is true for triangles, too. We can use triangles to build all kinds of structures as bridges. And we can take a polygon as a collection of triangles. That is to say that a polygon can be partitioned into many triangles.

Fig. 2.0

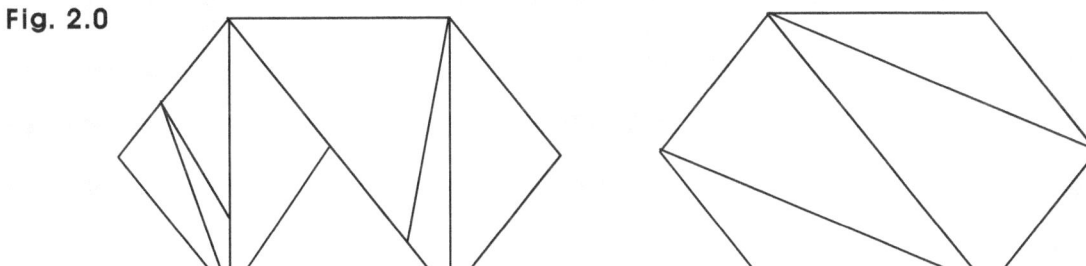

And also, we can take a triangle as a collection of many triangles called right triangles.

So a right triangle is a basic triangle, and has an angle called a right angle, which is 90°.

And the side facing the right angle is called the hypotenuse, and the other two sides perpendicular to each other are called the legs.

We call such a triangle a right triangle, since it has a right angle. So showing a right triangle, we want to show where the right angle is. Usually, we place a small rectangle or square at the vertex that has the right angle, 90°.

Fig. 2.1

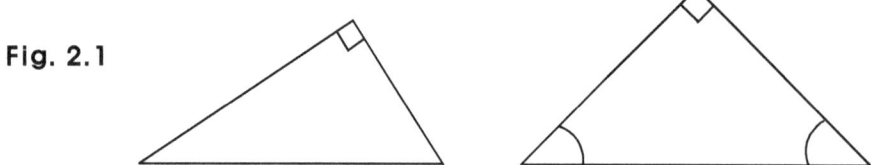

And we can partition a right triangle into many other right triangles, too.

Fig. 2.2

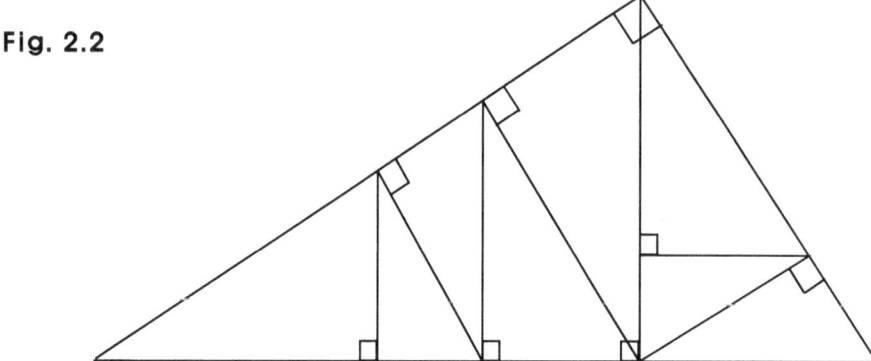

So we can keep generating as many right triangles as we want from a right triangle.
And of course, we can construct as many right triangles as we want inside any triangle.

Fig. 2.3

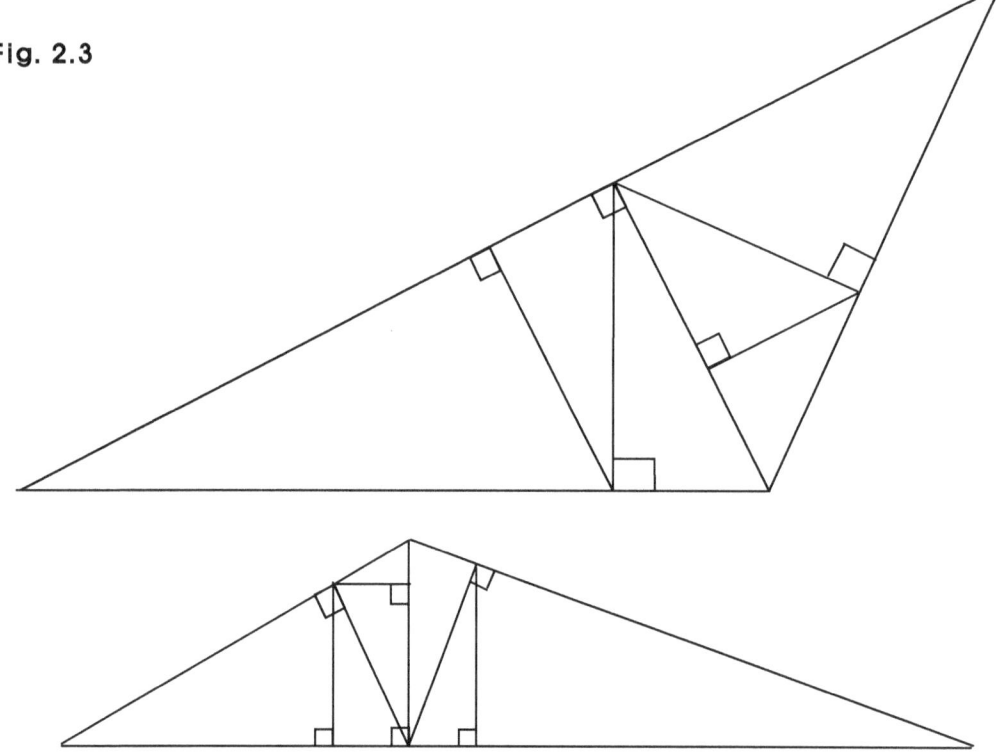

And using a right triangle, we can define an important tool in math, and the tool is called a slope. What slope?

It indicates how a line or a line segment is slanted or inclined. That is, it shows the degree of inclination of a line or a line segment. And we can specify the degree, that is, the slope by means of the ratio between the legs of a right triangle. What ratio though?

The slope is the ratio of the vertical leg to the horizontal leg. That is to say that the ratio is: the vertical side over the horizontal side, in a right triangle, of course.

Fig. 2.4 Uphill

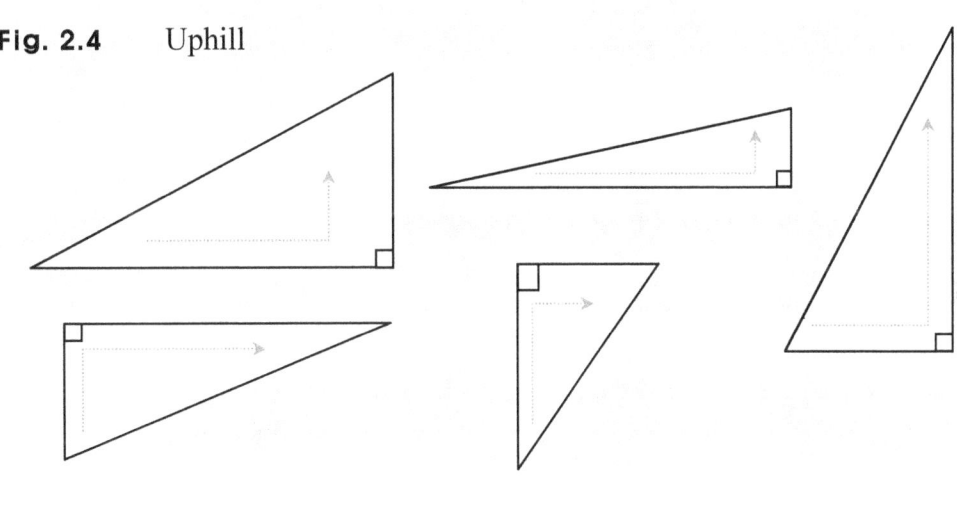

Fig. 2.5 Downhill

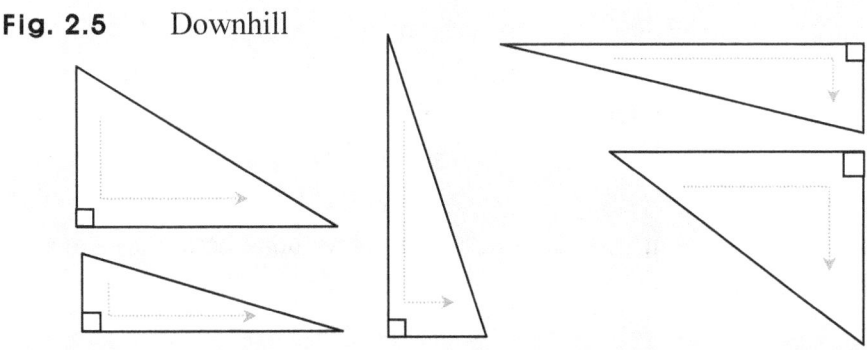

So the ratio called the slope can show the steepness, that is, how steep a right triangle is.

And specifically, the slope specifies the steepness of the hypotenuse in a right triangle.

What then, is it for?

In the case of uphill, in each right triangle, we take the vertical side for a positive change. That is, we give a positive value to the vertical side.

In the case of downhill though, we take the vertical side for a negative change. That is, we give a negative value to the vertical side.

And in both cases, we take the horizontal side for a positive change. So we give a positive value to the horizontal side.

And we know that the slope is: the vertical side over the horizontal side.

So in the case of uphill, the slope in each right triangle is positive.

In the case of downhill however, the slope in each right triangle is negative.

And thus, we can use the idea of the slope to explain how fast or slow things change as other things change. So in that case, we should call it a rate rather than just a ratio.

More specifically, we can call it *a rate of change*.

For instance, if a volume changes as time changes, we can call it a time rate of change in volume. So it shows how fast (or slow) the volume changes as time changes, and we can call it a speed. More precisely though, we call it a velocity, the magnitude of which is a speed. So it shows how fast or slow the volume increases or decreases as time increases.

And such an idea called the slope, that is, a rate of change is the fundamental idea in a very useful math called *calculus*.

And we know that the idea above is from right triangles. So right triangles are important.

It's not just important though. We cannot do much without it in math.

Is that all then, we can get from a right triangle?

We can get other crucial ideas from right triangles.

One of them is the <u>distance formula</u>, often called Pythagorean theorem, too.
And the formula shows the fact below:

The square of the distance between the two endpoints of the hypotenuse is the sum of the square of one leg and the square of the other leg.

In short, the square of the hypotenuse is the sum of the squares of the two legs.

So assuming h is the hypotenuse, a and b are the two legs, we get: $h^2 = a^2 + b^2$.

And some other ideas are <u>trigonometric ratios</u>, called trig-ratios, for short.
They are called the sine, the cosine, and the tangent.

The sine is the ratio of the vertical leg to the hypotenuse.
So if the sine is 0.5 and the hypotenuse is 3, the vertical leg is 1.5, which is the product of the sine and the hypotenuse: 1.5 = 0.3 x 3.

We can thus, put the vertical leg the way below:

The vertical leg is the product of the sine and the hypotenuse. That is, the vertical leg is the sine times the hypotenuse. So multiplying by the sine, we get the leg vertical.
So what is the sine about?

The sine is about the vertical leg, often just called *the opposite*.

The cosine is the ratio of the horizontal leg to the hypotenuse.
So if the cosine is 0.2, and the hypotenuse is 5, the horizontal leg is 1, which is the product of the cosine and the hypotenuse.

We can therefore, put the horizontal leg the way as follows:

The horizontal leg is the product of the cosine and the hypotenuse. That is, the horizontal leg is the cosine times the hypotenuse. So multiplying by the cosine, we get the leg horizontal. So what is the cosine about?

The cosine is about the horizontal leg, often just called *the adjacent*.

And the tangent is the ratio of vertical leg to the horizontal leg, which means the slope of the right triangle. So the tangent is the slope, often called the rise-over-run.

And since the tangent is the ratio of the vertical leg to the horizontal leg, we can say that the tangent is the vertical leg over the horizontal leg.
So given the tangent and the horizontal leg, we can get the vertical leg.   How?

Multiplying the horizontal leg by the tangent, we get the vertical leg.

What if we are given the tangent and the vertical leg, and want to find the horizontal leg?

We can get the horizontal leg dividing the vertical leg by the tangent, or multiplying the vertical leg by the reciprocal of the tangent. And the reciprocal is called the cotangent.

And those trig-ratios form a math area called trigonometry.

So right triangles are places where many important math tools get made.

3. Ratios from Right Triangles

We have another important fact about a right triangle, and the fact is as follows:

• A right triangle is the place where a special geometry called trigonometry begins.

From a right triangle, we can get another important tool called a *trigonometric ratio*, called a *trig-ratio*, for short. We can get in fact, six of those in kinds.
Getting such a ratio, we use two of the three sides in a right triangle. So a trigonometric ratio is basically made of two sides in a right triangle. What are the two sides though?

To begin with, we use different names for the legs of a right triangle.
And naming them, we consider a particular angle we refer to when getting a trig-ratio.

One of the two is called the adjacent side called briefly the <u>adjacent</u>, because it is adjacent to the particular angle stated above. And the other is called the opposite side called just the <u>opposite</u>, because it is opposite of the particular angle.

So in trigonometry, we put a right triangle the way below:

Fig. 3.0

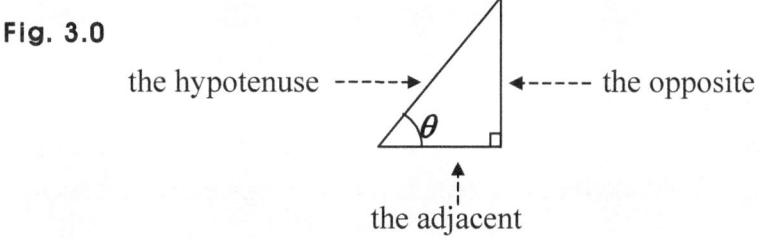

the hypotenuse ------>   <----- the opposite

θ

the adjacent

In the figure above, θ read as theta is the particular angle in the case above.
Depending on the particular angle θ, we get a different ratio, so the angle θ is crucial.

We have three basic trig-ratios, each of which has its multiplicative inverse, often called the reciprocal, too. So multiplying each of the three and its reciprocal, we get 1.

We want to note however, that the inverse of a trig-ratio is an angle, and that the multiplicative inverse or the reciprocal of a trig-ratio is another trig-ratio, which is a number. What are those three basic trig-ratios, then?

Getting a trigonometric ratio, we pick two of the three sides in a right triangle, and then, take a ratio between the two. We don't just pick two sides though, and do not simply take a ratio between the two.

Two of the three sides make an important angle in a right triangle. And we may want to call the important angle the governing angle. That's because such an angle governs or determines the ratios. What sides are the two though?

Assuming first, in a right triangle, *H* is the hypotenuse, *A* is the adjacent, and *O* is the opposite, we can put the right triangle the way below:

Fig. 3.1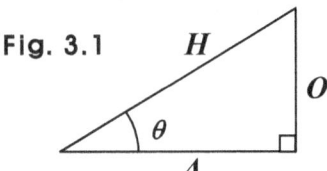

Then, we call *θ* (read as theta) the *governing angle*, because we have chosen the side *A* to be the adjacent, or we have chosen the side *O* to be the opposite. So the adjacent is the side adjacent to the governing angle, and the opposite is the side opposite of (or facing) the governing angle.

And thus, the two sides are the adjacent and the hypotenuse. So in a right triangle, <u>the adjacent and the hypotenuse make the governing angle.</u>

And in return, *the governing angle determines every trig-ratio* in the right triangle.

Depending on the way we look at the right triangle though, either of the two sides other than the hypotenuse can be the adjacent. So using a wrong side as the adjacent, we get wrong ratios. Thus in short, *the wrong adjacent makes wrong ratios.*

So it is crucial to <u>choose the adjacent correctly</u>.

Now, we can get six trig-ratios in kind from a right triangle, and each is between two of the three sides. What then, are the six ratios?

Two are between the opposite and the hypotenuse, another two are between the adjacent and the hypotenuse, and the other two are between the opposite and the adjacent.

Of the six though, three are in fact, the reciprocals of the other three, which are called therefore, three basic trig-ratios. And more specifically, of the three basic ratios:

- One is called *sine*, which is denoted by **sin**, and is the ratio of the opposite to the hypotenuse, that is, the opposite over the hypotenuse: the opposite / the hypotenuse.

- Another is called *cosine*, which is denoted by **cos**, and is the ratio of the adjacent to the hypotenuse, that is, the adjacent over the hypotenuse: the adjacent / the hypotenuse.

- And the other is called *tangent*, denoted by **tan**, which is the ratio of the opposite to the adjacent, that is, the opposite over the adjacent: the opposite / the adjacent. So the tangent can tell us the slope of the right triangle.

Suppose now, the governing angle is θ, read as theta, which is the eighth letter of the Greek alphabet. Then, we put the three basic trig-ratios the way below:

• To begin with, the sine of the governing angle θ is: **sin θ**, and is the ratio of the opposite to the hypotenuse, so **sin θ** is: the opposite over the hypotenuse.

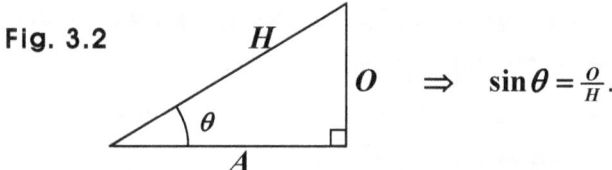

Fig. 3.2 $\quad \Rightarrow \quad \sin\theta = \frac{O}{H}.$

And we read **sin θ** as: sine of θ or just sine θ for short. So for instance, we read **sin 30°** as sine of 30° or just sine 30°, which equals: **sin π/6**, since 30° is **π/6** in radian.

• Next, the cosine of the governing angle θ is **cos θ**, and is the ratio of the adjacent to the hypotenuse, so **cos θ** is: the adjacent over the hypotenuse.

Fig. 3.3 $\quad \Rightarrow \quad \cos\theta = \frac{A}{H}.$

And we read **cos θ** as: cosine of θ or just cosine θ for short. So for instance:
We read **cos 45°** as cosine of 45° or just cosine 45°, which equals: **cos π/4**.

• And next, the tangent of the governing angle θ is **tan θ**, and is the ratio of the opposite to the adjacent, so **tan θ** is: the opposite over the adjacent.

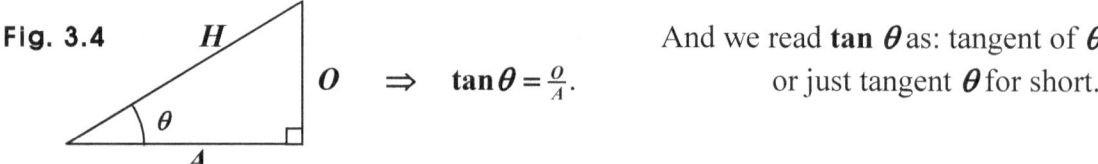

Fig. 3.4 $\quad \Rightarrow \quad \tan\theta = \frac{O}{A}.$

And we read **tan θ** as: tangent of θ or just tangent θ for short.

So for instance, we read **tan 60°** as tangent of 60° or just tan of 60°, which is equal to: **tan π/3**. And we can call the tangent the slope of the hypotenuse, because the tangent is: the opposite over the adjacent.

And thus, putting threads together, and assuming A is the adjacent in the right triangle below, we have to use θ as the governing angle, and can put the three basic trig-ratios the way as follows:

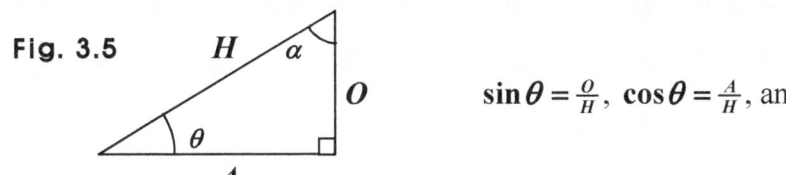

Fig. 3.5

$$\sin\theta = \tfrac{O}{H}, \ \cos\theta = \tfrac{A}{H}, \text{ and } \tan\theta = \tfrac{O}{A}.$$

And for another instance, assuming β is the governing angle in the right triangle below, we use q as the adjacent, and can put the three basic trig-ratios the way below:

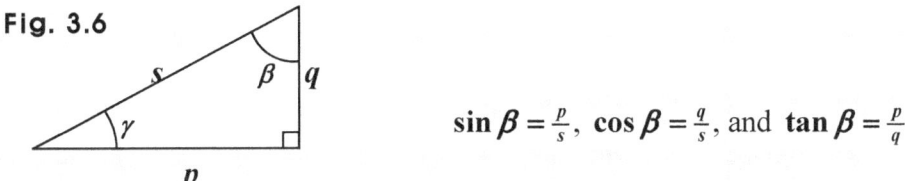

Fig. 3.6

$$\sin\beta = \tfrac{p}{s}, \ \cos\beta = \tfrac{q}{s}, \text{ and } \tan\beta = \tfrac{p}{q}.$$

And we have three popular governing angles, which are $30°$, $45°$, and $60°$. That's because, we can easily get the trig-ratios for those angles using two triangles isosceles.

One is a regular (equilateral) triangle, where every angle is $60°$, and the other is a right triangle isosceles, where two of the three angles are equal, and thus, are $45°$ each.

So to begin with, cutting in half a regular triangle, we can get a right triangle as below:

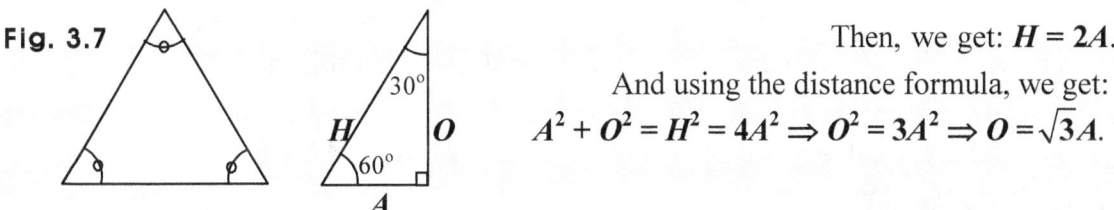

Fig. 3.7

Then, we get: $H = 2A$.

And using the distance formula, we get:

$$A^2 + O^2 = H^2 = 4A^2 \Rightarrow O^2 = 3A^2 \Rightarrow O = \sqrt{3}A.$$

So we get:

Fig. 3.8

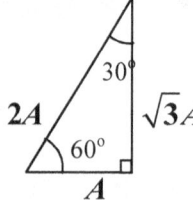

And thus, we can readily get:

sin 60° = the opposite / the hypotenuse $=\frac{\sqrt{3}}{2}$. **sin 30°** = the opposite / the hypotenuse $=\frac{1}{2}$.

cos 60° = the adjacent / the hypotenuse $=\frac{1}{2}$. **cos 30°** = the adjacent / the hypotenuse $=\frac{\sqrt{3}}{2}$.

tan 60° = the opposite / the adjacent $=\frac{\sqrt{3}}{1}$. **tan 30°** = the opposite / the adjacent $=\frac{1}{\sqrt{3}}=\frac{\sqrt{3}}{3}$.

And next, assuming *A* is the adjacent of a right triangle isosceles, we can put the triangle the way below:

Fig. 3.9

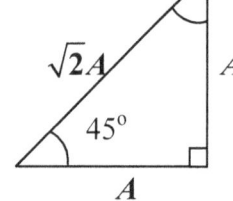

So we get:

sin 45° = the opposite / the hypotenuse $=\frac{1}{\sqrt{2}}=\frac{\sqrt{2}}{2}$.

cos 45° = the adjacent / the hypotenuse $=\frac{1}{\sqrt{2}}=\frac{\sqrt{2}}{2}$.

tan 45° = the opposite / the adjacent $=\frac{1}{1}=\mathbf{1}$.

Is it the case though, for instance, we get: **sin 30° = 1/2** in every right triangle where the governing angle is 30°?

Yes, it is. So for instance, we have: **cos 60° = 1/2** in every right triangle where the governing angle is 60°. And the same is true for all the other trig-ratios, too.

And the full explanations will be covered in the book, **ALGEBRA EXAMPLES TRIGONOMETRY**.

₄. Identical or Similar?

To begin with, among triangles, we have three in kinds or categories.
One is scalene, another is isosceles, and the other is regular or equilateral.
And of the three above, two can be called symmetric.
What then, are the two?

The two kinds are isosceles and regular. So triangles isosceles or regular are symmetric.
We know however, a regular triangle can be called an isosceles triangle, because an
isosceles triangle has two sides equal, and a regular triangle has three sides equal, so
anyway, two sides are equal in a regular triangle. And thus, triangles isosceles include
regular triangles.

So if a triangle has two sides equal, it is isosceles, and is symmetric. And we know if a
triangle is isosceles, it has two angles equal, too. So if a triangle has two angles equal, it
is isosceles, and is symmetric, also. Symmetric about what though?

If a triangle is isosceles, it is symmetric about a line that passes through a vertex and is
perpendicular to the side connecting the two sides equal. And we call the line the *axis of
symmetry.*

Fig. 4.0

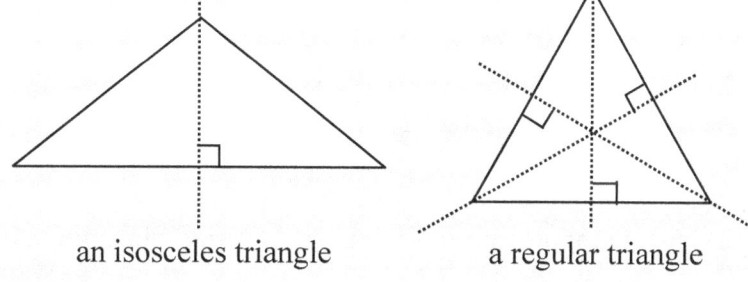

an isosceles triangle   a regular triangle

So we can see that cutting an isosceles or regular triangle along the axis of symmetry, we get two right triangles identical.

Suppose next, three line segments can make a triangle.

Then, how many different triangles can we make using the three?

We can make one triangle only. That is to say that if two triangles share the same set of three sides, the two triangles are identical. How come?

Suppose we made a triangle connecting three rods, and connecting them, we used pins so that the connections are lose, that is, no friction exists between each pin and each rod.

Fig. 4.1

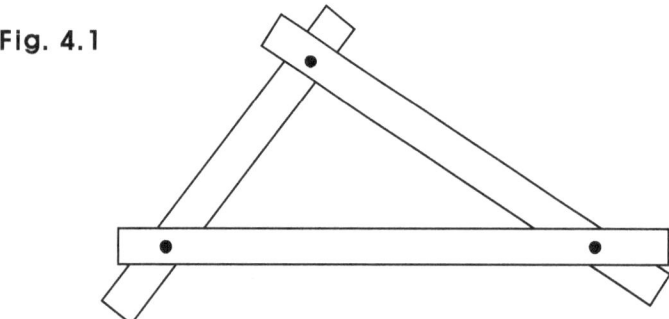

Can we then, get the triangle changed, without moving any of the positions of the pins, of course?

None of the rods can move as if they were riveted or glued together.
In other words, we cannot get any of the rods moved once the rods have been pinned together making a triangle. So we cannot make different triangles using a set of three line segments.

What then, about making a tetragon connecting four rods, using pins, of course?

Fig. 4.2

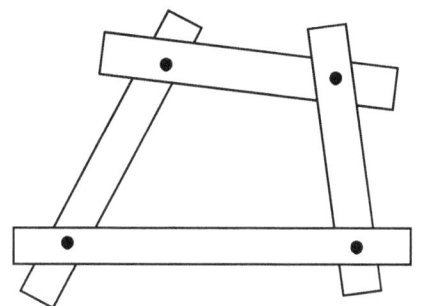 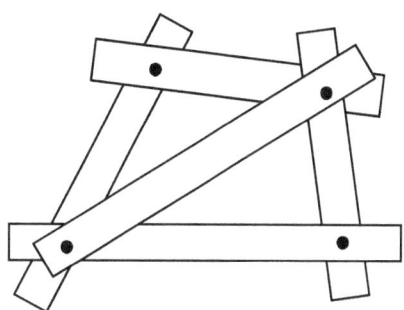

Making a tetragon like the one on the left in the figure above, we can get it changed even if we push slightly any of the rods. So we can make many different tetragons using a set of four line segments. Making however, an object like the one on the right in the figure above, we cannot get it changed if we do not move any of the positions of the pins.

That's because it is partitioned into triangles, and if three line segments can make a triangle, we can make one triangle only. That is to say that if two triangles share the same set of three sides, the two triangles are identical.

What then, about using three angles, if they can make a triangle, of course? For instance, using three angles 30°, 60°, and 90°, can we make only one triangle?

We can make as many triangles as we want. In each of the triangles, the three angles are 30°, 60°, and 90°, of course.
Those triangles do not share though, the same set of three sides. And we call all those triangles similar triangles.
So similar triangles share the same set of three angles, but do not share the same set of three sides. For instance, if two triangles are similar, and one is made of 35°, 60°, and 85°, the other has to be made of 35°, 60°, and 85°, too. And thus, putting threads together, we can say that:

• If two triangles are *identical*, they share the same set of three sides, and of course, share the same set of three angles, too.

• And if two triangles are *similar*, they share the same set of three angles, but do *not* share the same set of three sides.

Suppose for instance, the two triangles below are identical, that is: $\triangle ABC \equiv \triangle DEF$:

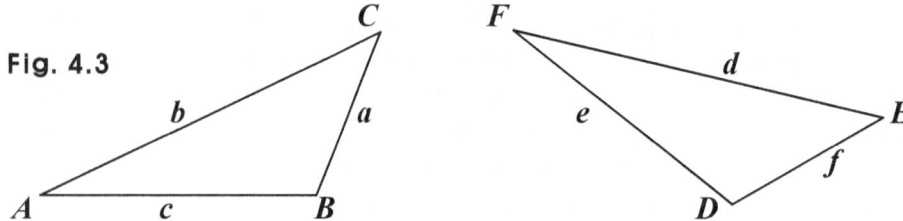

Fig. 4.3

Then, we get: $b = d$, $c = e$, and $a = f$.
And of course, we get this, too: $\angle A = \angle F$, $\angle B = \angle D$, and $\angle C = \angle E$.

So checking to see if two triangles are the same, that is, identical, do we have to actually compare all the six sides?

Not necessarily.

If two triangles share the same set of two sides, and the angle between the two sides in one triangle is the same as the angle between the two sides in the other triangle, then the two triangles are identical.
For instance, in the two triangles $\triangle ABC$ and $\triangle DEF$ above, if $b = d$, $c = e$, and $\angle A = \angle F$, we get: $\triangle ABC \equiv \triangle DEF$.

Thus for instance, the two triangles below are identical:

Fig. 4.4

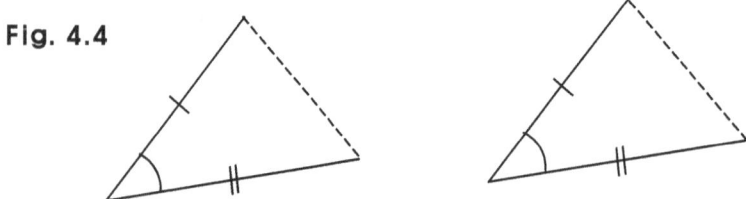

And we can check to see if two triangles are the same the way below, too:

If two triangles share the same set of two angles, and the side between the two angles in one triangle is the same as the side between the two angles in the other triangle, then the two triangles are identical. For instance, in the two triangles $\triangle ABC$ and $\triangle DEF$ above, if $\angle A = \angle F$, $\angle B = \angle D$, and $c = e$, we get: $\triangle ABC \equiv \triangle DEF$.

Fig. 4.5

• And checking to see if two triangles are similar, we do not have to actually compare all the six angles.

If two triangles share the same set of two angles, they have to share the same set of three angles, and thus, are similar.

That's because the sum of the three angles in every triangle is $180°$, which means, the sum is the same. For instance, if two angles in a triangle are $20°$ and $60°$, and another triangle has the two angles, too, the two triangles are similar, because both triangles have to have $100°$ each, too.

• And technically, identical triangles are similar triangles, too, because identical triangles share the same set of three angles. Normally though, similar triangles are different triangles. Saying therefore, similar triangles, we mean different triangles.

What do we mean by though, similar triangles?

They don't just look similar, of course. Though similar triangles are different triangles, they are not just randomly different. In fact, they have to a couple of things the same.

First, they have to share the same set of three angles. And next, if two triangles are similar, the ratio between corresponding sides has to be the same.
That is to say that all the three pairs of corresponding sides share the same ratio.
What do we mean by though, corresponding sides?

They are the sides facing the same angle.

Suppose for instance, in the two triangles below, $\angle A = \angle D$, $\angle B = \angle E$, and $\angle C = \angle F$.

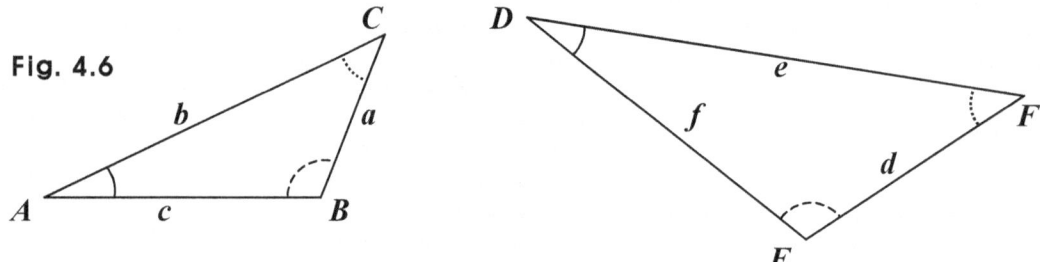

Fig. 4.6

Then, we get: $\triangle ABC \approx \triangle DEF$, which means the two triangles are similar.

And we can say that b corresponds to e, c corresponds to f, and a corresponds to d.

So we get: $b : e = c : f = a : d$.   That is to say that we get: $\frac{b}{e} = \frac{c}{f} = \frac{a}{d}$.

So for instance, if $b : e = 2 : 3$, we get: $b/e = c/f = a/d = 2/3$.

And in fact, we can put the two triangles $\triangle ABC$ and $\triangle DEF$ above the way below, too:

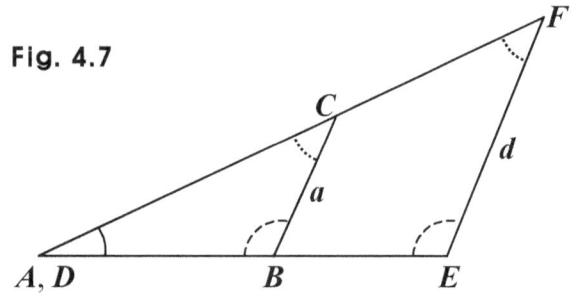

Fig. 4.7

Then, of course, a is parallel to d.
So both triangles share
the same set of three angles.

And for instance, in the figure above, assuming $a = 5$, $AC = 10$, and $d = 8$, and finding the length of CF, we can get it the way as follows:

Finding AF first, we can get CF, because $CF = AF - AC$.   So finding AF now, we get:

$\frac{AC}{AF} = \frac{a}{d} \Rightarrow \frac{10}{AF} = \frac{5}{8} \Rightarrow 10 = AF \cdot \frac{5}{8} \Rightarrow AF = 10 \cdot \frac{8}{5} = 16$.   Thus, we get: $CF = AF - AC = 6$.

And for another instance, suppose in the figure below, we have:

AD // LE, *AG // BF*, and *CJ // DG*. (Note that *AD // LE* means *AD* is parallel to *LE*.)

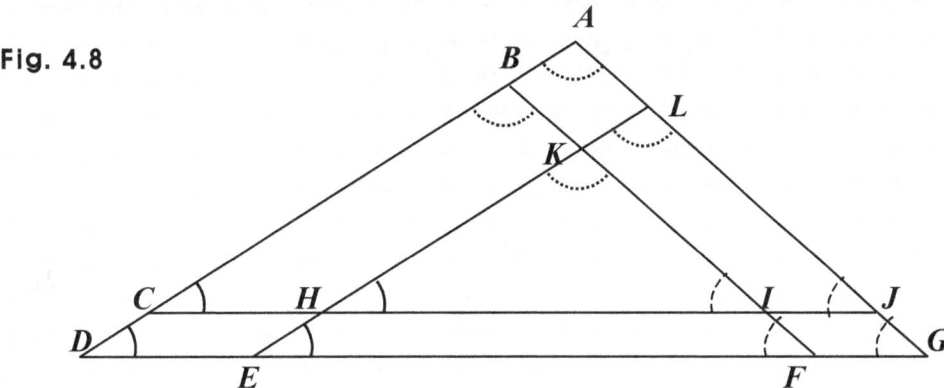

Fig. 4.8

Then, △*ADG*, △*BDF*, △*LEG*, △*LHJ*, △*KEF*, △*KHI* . . . are all similar triangles. And thus, all the corresponding angles are the same. That is to say that we get:

∠*BCH* = ∠*CDE* = ∠*HEF* = ∠*KHI*.
∠*CBK* = ∠*BAL* = ∠*HKI* = ∠*KLJ*.
∠*KIH* = ∠*IFE* = ∠*JGF* = ∠*LJI*.

And the ratio between each pair of corresponding sides is the same, too.
So for instance, considering △*ADG* and △*KHI*, we can get: $\frac{KH}{AD} = \frac{HI}{DG} = \frac{KI}{AG}$.

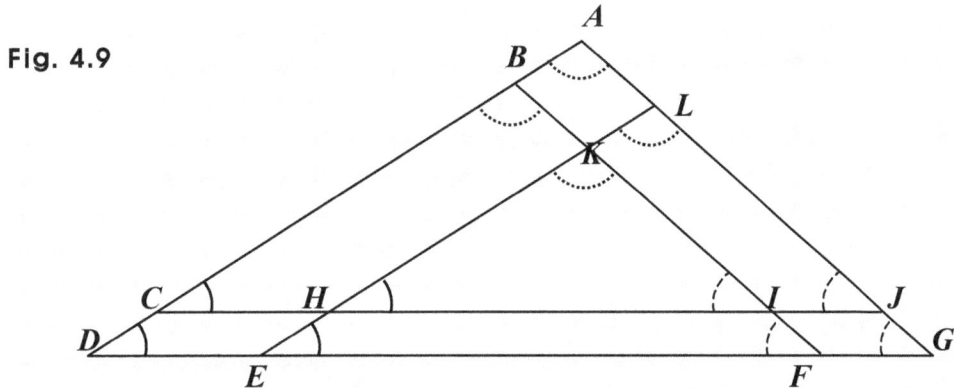

Fig. 4.9

Next, looking at △*ADG* and △*KEF*, we can get: $\frac{AD}{KE} = \frac{DG}{EF} = \frac{AG}{KF}$.
And next, looking at △*BCI* and △*LEG*, we can get: $\frac{BC}{LE} = \frac{CI}{EG} = \frac{BI}{LG}$.

• And also, the ratio between two sides in one triangle is the same as the ratio between the corresponding two sides in another triangle similar to the one.

For instance, in the two triangles $\triangle ADG$ and $\triangle KHI$, the ratio of **AD** to **DG** is the same as the ratio of **KH** to **HI**. That is, we get: $\frac{AD}{DG} = \frac{KH}{HI}$. How come?

Suppose the length of **AD** is twice the length of **KH**.
Then, since $\triangle ADG$ and $\triangle KHI$ are similar, the length of **DG** is twice the length of **HI**, too, and also, the length of **AG** is twice the length of **KI**.

Thus, we get: $\frac{AD}{DG} = \frac{2KH}{2HI} = \frac{KH}{HI}$. And the same is true for the other pairs, too.

That is, we get: $\frac{AD}{AG} = \frac{KH}{KI}$, and $\frac{AG}{GD} = \frac{KI}{IH}$.

And of course, the same is true, too, for all the other similar triangles.
So let's, for instance, find the length of **EG** in the figure below assuming **BC = 2**, **CI = 3**, and **LE = 4**.

Fig. 4.8

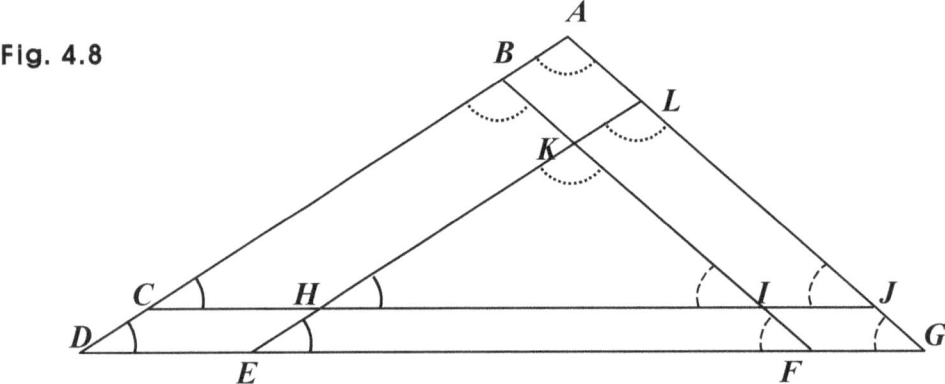

Then, $\triangle BCI$ and $\triangle LEG$ are similar triangles.

So we can get: $\frac{BC}{CI} = \frac{LE}{EG} \Rightarrow \frac{2}{3} = \frac{4}{EG} \Rightarrow 2EG = 12 \Rightarrow EG = 6$.

5. Areas of Triangles

Taking the magnitude of a line segment, we take the length of it.
And taking the magnitude of a triangle, we take the area of it.
How then, can we take the area of a triangle?

Taking, for instance, the area of a rectangle, we measure first, its two dimensions. And we call the two dimensions the length and the width or the base and the height. And then, taking the product of the two, we get the area. So given its dimensions, we can get the area taking the product of the dimensions. What then, about the area of a triangle?

We can make a polygon putting together triangles, and can make a rectangle putting together two same right triangles. What then, are the two legs in each right triangle?

The two legs are the base and the height, that is, the two dimensions of the rectangle. And we know that the area of the rectangle is the product of its two dimensions. So taking the half of the product, we get the area of the right triangle.

And thus, taking the area of a right triangle, we take the product of the two legs, and then, divide it by 2, that is, multiply it by one half, 1/2.

What if however, the triangle is not a right triangle?

We can get the area of such a triangle, too, using the way we used above.
This time though, we use the way we use getting the area of a parallelogram.

Taking the area of a parallelogram, we measure first, its two dimensions, which are the base and the height. And then, taking the product of the two, we get the area. So given its base and height, we can get the area taking the product of the base and the height. What do we mean by though, the base and the height in a parallelogram?

We can in fact, take as the base any of the four sides. Then, the height is the distance between the base and the side facing the base.
So for instance, putting a parallelogram *ABCD* the way below, we can take the side *AB* as the base, and can take as the height the distance between the two sides *AB* and *CD*.

Fig. 5.0

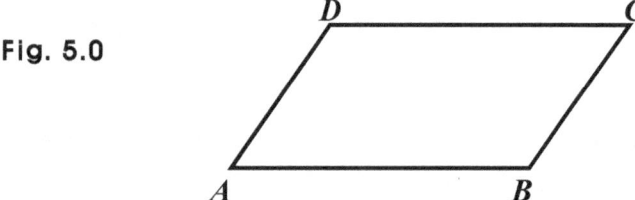

And indicating a parallelogram, we often use a symbol, which is a small parallelogram. So calling the height is *H*, we can put ▱*ABCD* the way below:

Fig. 5.1

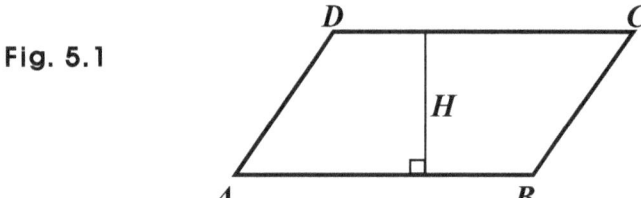

How come then, is the product of the base and the height is the area?

We can partition the parallelogram above so that we get a rectangle and two same right triangles the way below:

Fig. 5.2

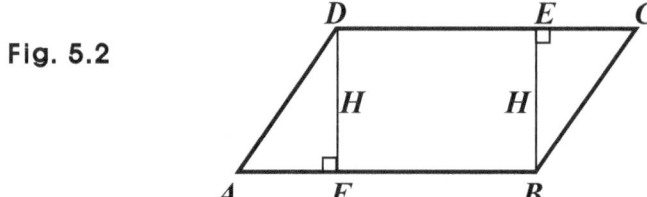

Then, moving the right triangle on the left next to the one on the right, we get a rectangle.

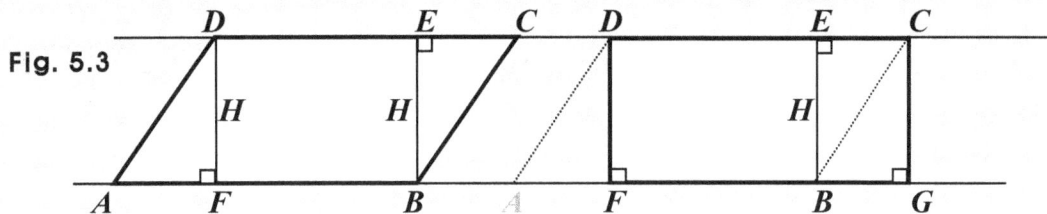

Fig. 5.3

Then, we get: **AB = FG**.

So the area of ▱**ABCD** is the same as the area of the rectangle **FGCD**.

And thus, the area of ▱**ABCD** is the product of the base **AB** and the height **H**.

What then, about the area of a triangle?

We can make a parallelogram putting together two same triangles.

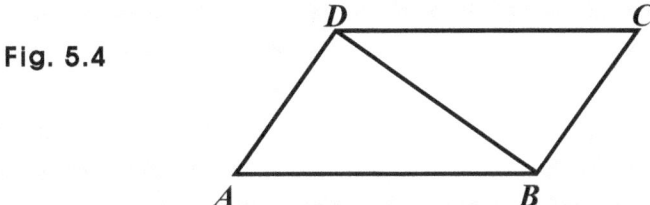

Fig. 5.4

The two triangles **ABD** and **CDB** are the same. That is, we have: **△ABD ≡ △CDB**.

So the area of ▱**ABCD** is twice the area of **△ABD**.

And we know that the area of a parallelogram is the product of the base and the height, and the height in the case above is the distance between **AB** and **CD**.

So taking the half of the product, we get the area of the triangle.

And assuming the height is **H**, we can put ▱**ABCD** the way below:

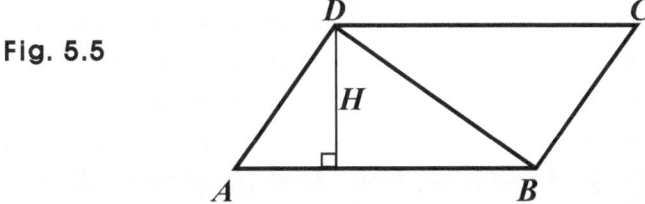

Fig. 5.5

We know **D** is the vertex facing the side **AB** in **△ABD**. So taking the area of a triangle, we take the product of one of its side and the distance from the side to the vertex facing the side, and then, divide the product by 2, that is, multiply it by one half, 1/2.

And the side chosen is called the *base* of the triangle, and the distance stated above is called the *height* of the triangle.

And thus, when taking the area of a triangle, we need to choose a side and take the side as the base, and we need to take the distance from the base to the vertex facing the base, and take the distance as the height.

And then, taking the product of the base and the height, and dividing the product by 2, we get the area of the triangle.

So in short, *the area of a triangle* is *half the base times the height.*
What if we want to take **PQ** as the base in the triangle **PQR** below?

Fig. 5.6

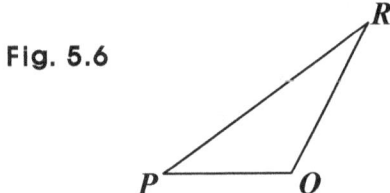

We know that the height is the distance from the base to the vertex facing the base. So assuming **h** is the height, we can take the height the way below:

Fig. 5.7

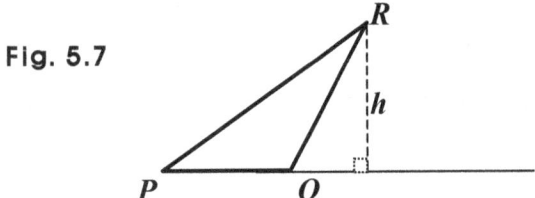

And putting together two of the same triangles as the one above, we can get:

Fig. 5.8

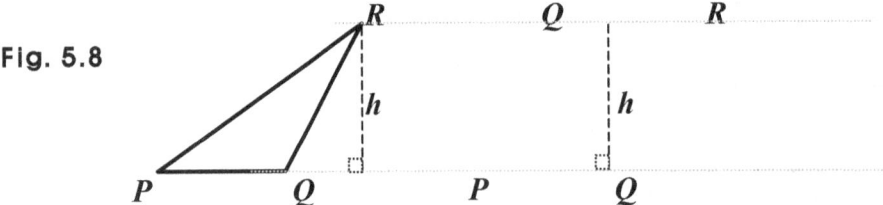

And thus, assuming the height is **h**, the base **PQ** is **b**, and the area is A, we get: $A = \frac{bh}{2}$.

Can we have though, different triangles sharing the same base and the same height?

Yes, we can. And we can say that different triangles sharing the same base and the same height have the same area. So for instance, all the triangles below have the same area.

Fig. 5.9

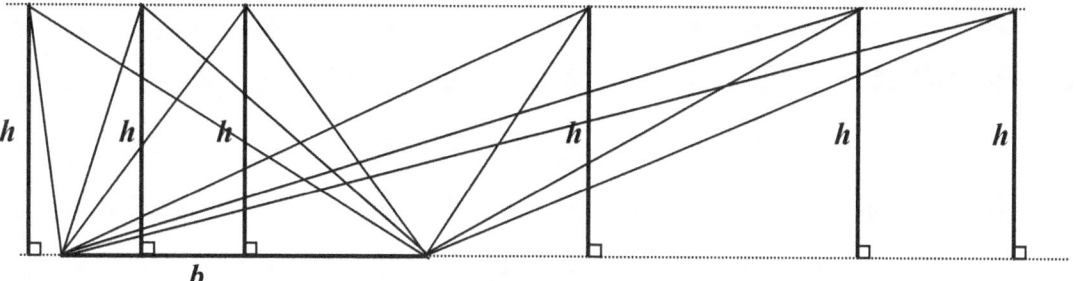

And we often use a symbol ⊥ to indicate objects are perpendicular to each other. So in the figure above, we have: $b \perp h$.

And taking the area of a triangle, we can take as the base any side of the triangle. Then, we can take as the height the distance from the base to the vertex facing the base. So for instance, we can take the base and its corresponding height any of the ways as follows:

Fig. 5.A

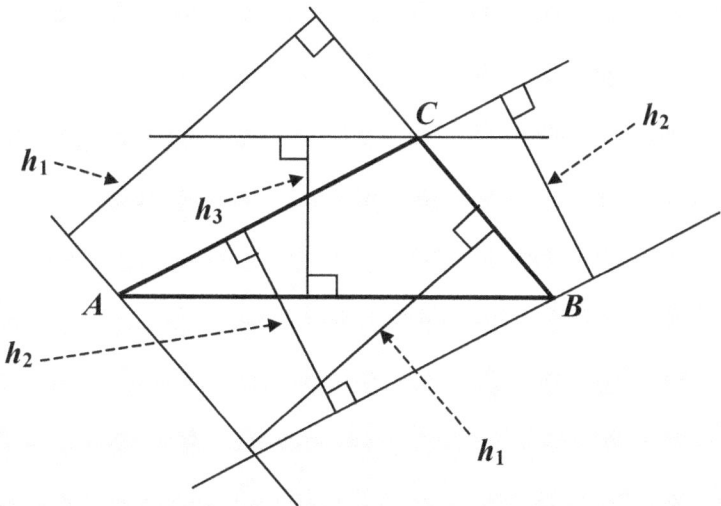

In the figure above, we have: $AB \perp h_3$, $BC \perp h_1$, and $CA \perp h_2$.

That is to say that:

The height h_3 corresponds to the base AB.
The base BC corresponds to the height h_1.
And the height h_2 corresponds to the base CA.

So assuming $AB = c$, $BC = a$, $CA = b$, and S is the area of $\triangle ABC$ above, we can put S the way below:

$$S = \frac{ah_1}{2} = \frac{bh_2}{2} = \frac{ch_3}{2}.$$

6. The Distance Formula

What is the distance formula?

We can use it to find the distance between two points.
And the distance is the length of the line segment connecting the two points, which are placed in a coordinate plane as the *x-y* plane or in a coordinate space as the *x-y-z* space.

And the formula is from a right triangle.
For instance, assuming *a* and *b* are the two legs in a right triangle, and *c* is the hypotenuse, and putting the three sides in the distance formula, we get: $a^2 + b^2 = c^2$. How come though, is the formula for the distance between two points?

Putting the right triangle above in the *x-y* plane, and assuming a point *(u, v)* is an end point of the hypotenuse, and another point *(s, t)* is at the other end, we can put the right triangle and the two points the way below:

Fig. 6.0

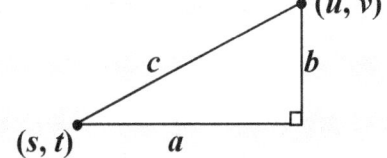

Then, the distance is the hypotenuse *c*.

And we often call it Pythagorean theorem, too.

The distance formula is one of the tools the most frequently used when we do math. So we cannot do much without it doing math. How do we know if it is the case though?

We can partition two same squares the way below:

Fig. 6.1

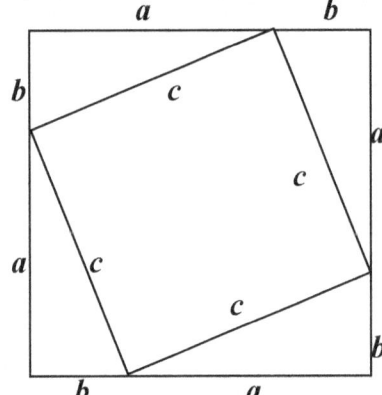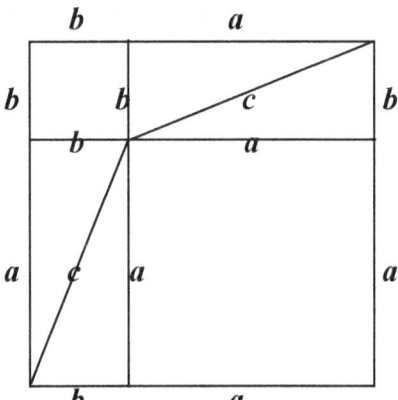

Each of the two squares has 4 right triangles, in each of which, *a* and *b* are the two legs, and *c* is the hypotenuse. And the square on the left has a smaller square *c* by *c*, and the square on the right has two smaller squares, one is *b* by *b*, and the other is *a* by *a*.

So the area of the square *c* by *c* is the same as the sum of the areas of the two squares, one of which is *b* by *b*, and the other is *a* by *a*. And thus, we get: $a^2 + b^2 = c^2$.

And we can put the idea above this way, too:

The area of the large square is: $(a + b)$ by $(a + b)$, and thus, is: $(a + b)^2$.

And the sum of the areas of the four right triangles is: $4 \cdot (ab/2) = 2ab$.

So subtracting the sum from the area of the large square, we get the area of the smaller square, which is: *c* by *c*, that is, c^2.

And thus, we get: $(a + b)^2 - 2ab = a^2 + 2ab + b^2 - 2ab = a^2 + b^2 = c^2$.

And we can put the same idea the way below, too.

To begin with, we can put three squares and a right triangle the way below.

Fig. 6.2

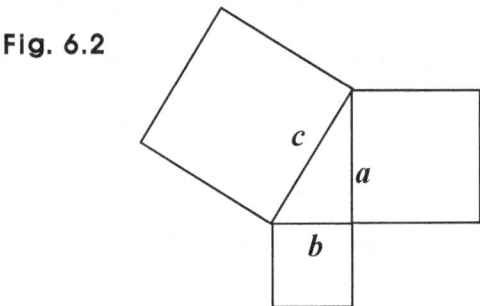

Suppose next, the square *c* by *c* is a square **ABCI**, the square *a* by *a* is a square **IFGH**, and the square *b* by *b* is a square **DEFC**.

Then, we want to show that the area of the square **ABCI** is the same as the sum of the areas of the two squares **CDEF** and **IFGH**.

Fig. 6.3

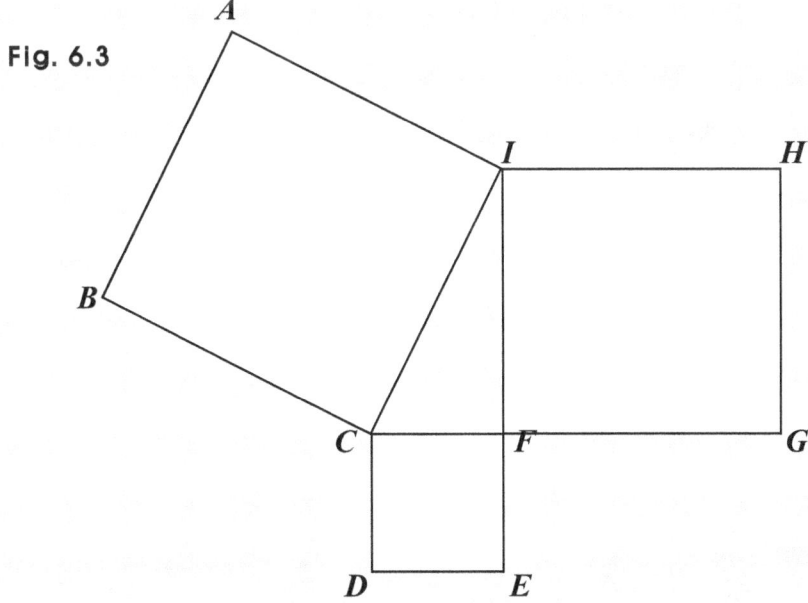

First, we can say that the area of **△ICH** is the same as the area of **△IGH**, which is the half the area of the square **IFGH**.

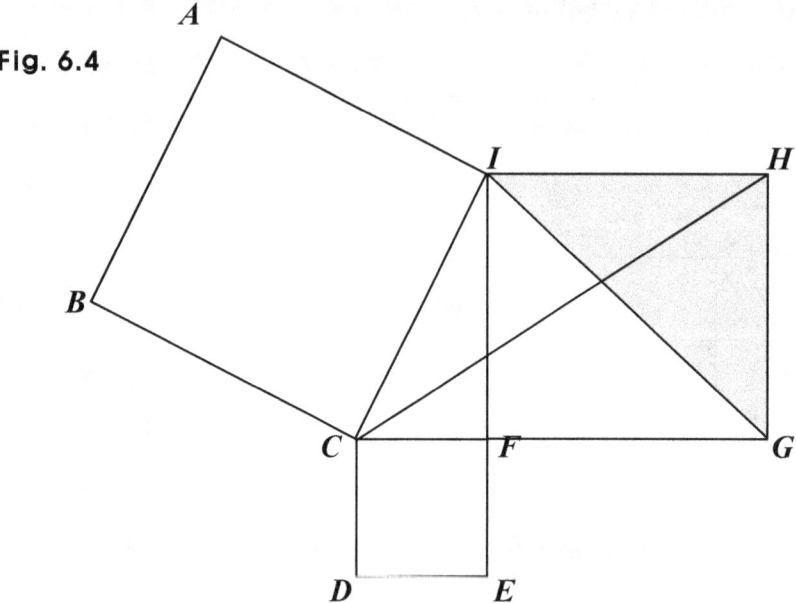

That's because assuming **IH** is the base, we can say that **HG** is the height of **ΔICH**, and also, is the height of **ΔIGH**.

Next, we can say that **ΔICH** is the same as **ΔIAF**. That is, both are identical.

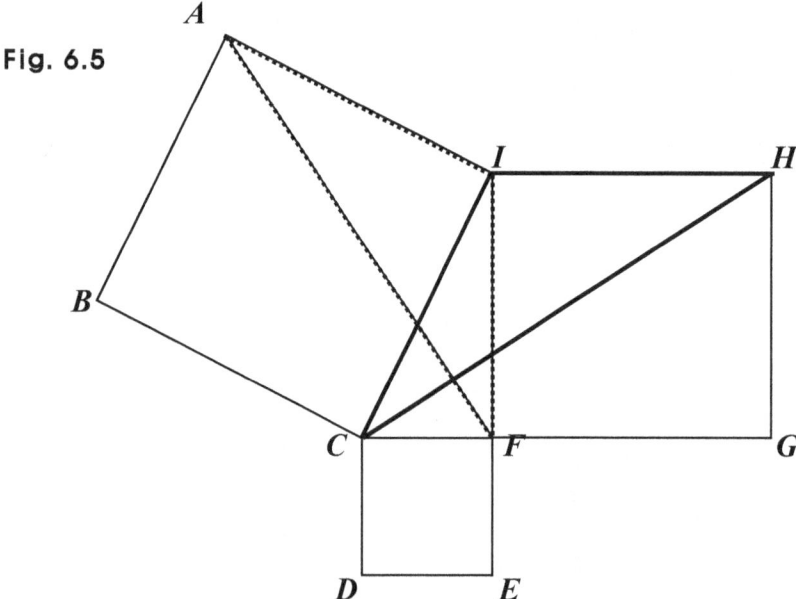

That's because **IH** is the same as **IF**, **IC** is the same as **IA**, and the angle **HIC** is the same as the angle **AIF**, and thus, **AF** is the same as **CH**. So we get: **ΔICH ≡ ΔIAF**.

Next, assuming a line **L** is parallel to the side **AI**, meets the side **IC** at **J**, and passes through the vertex **F**, we can say that the area of **ΔIAF** is the same as the area of **ΔIAJ**.

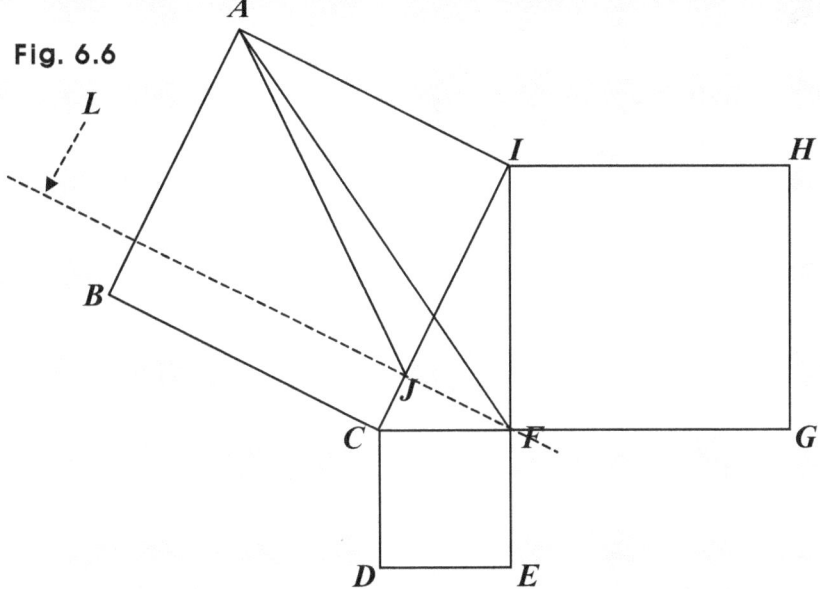

Fig. 6.6

It's because assuming **IA** is the base, we can say **IJ** is the height of both **ΔIAF** and **ΔIAJ**. Next, we can say that the area of **ΔICD** is the same as the area of **ΔFCD**, which is the half the area of the square **CDEF**.

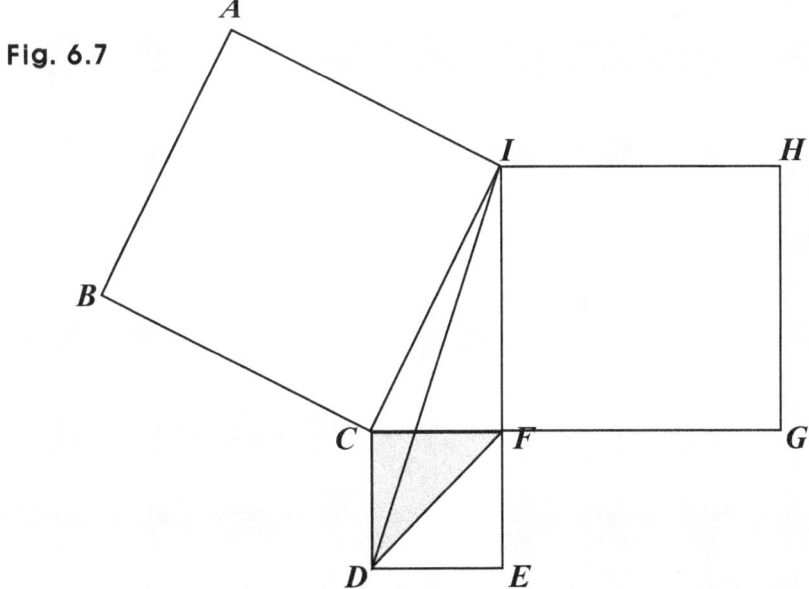

Fig. 6.7

That's because assuming **CD** is the base, we can say that **CF** is the height of **ΔICD**, and is the height of **ΔFCD**, too.

Next, we can say that **Δ*ICD*** is the same as **Δ*BCF***, and thus, both are identical.

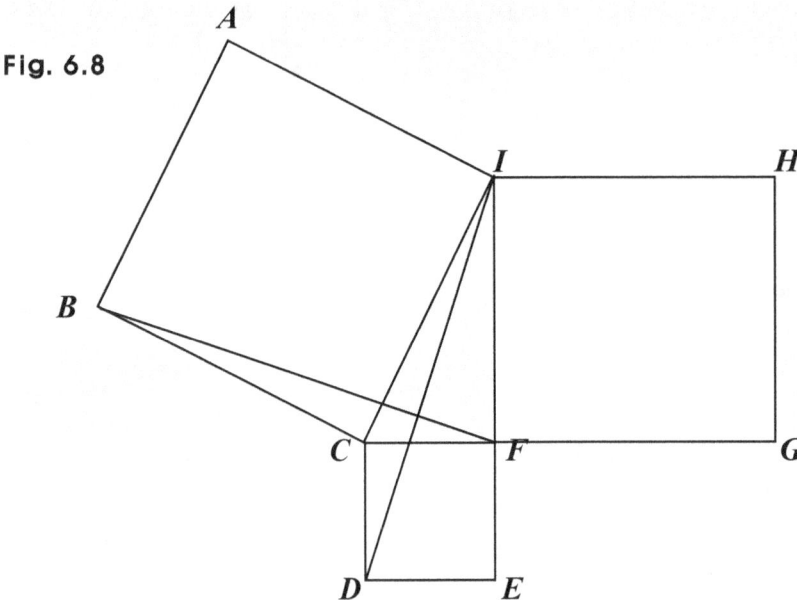

Fig. 6.8

That's because **IC** is the same as **BC**, **CD** is the same as **CF**, and the angle **BCF** is the same as the angle **ICD**, and thus, **BF** is the same as **ID**. So we get: **Δ*ICD* ≡ Δ*BCF***. Next, we know the line **L** is parallel to the side **AI**, meets the side **IC** at **J**, and passes through the vertex **F**. So we can say the area of **Δ*BCF*** is the same as the area of **Δ*BCJ***.

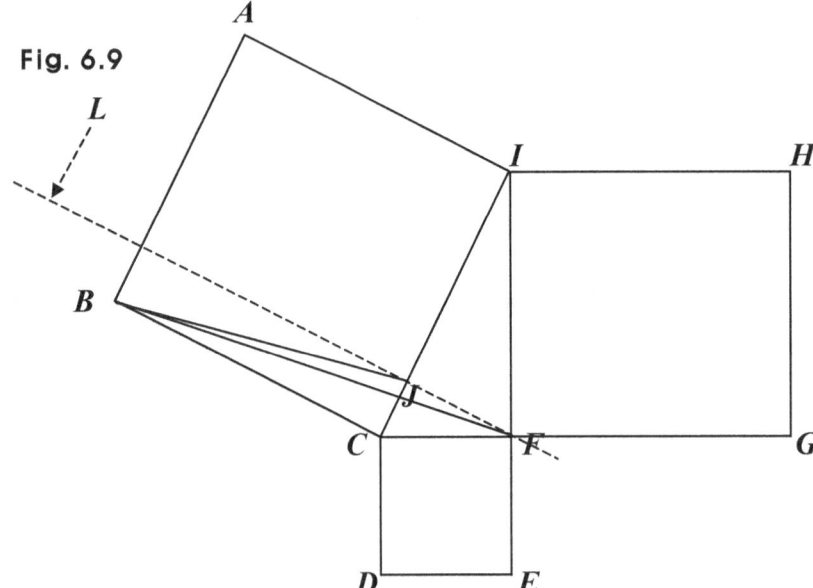

Fig. 6.9

That's because assuming **BC** is the base, we can say that **JC** is the height of **Δ*BCF***, and is the height of **Δ*BCJ***, too. Now, putting threads together, we have:

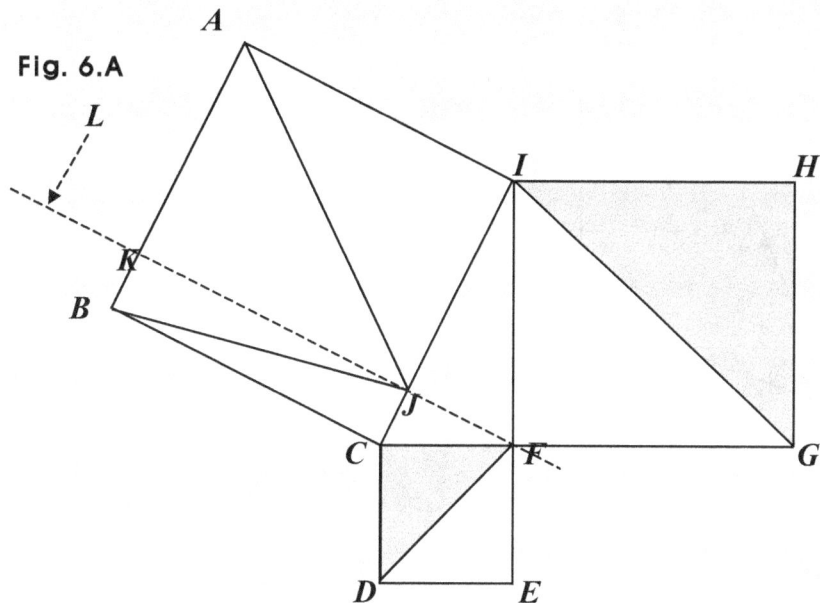

Fig. 6.A

Then, to begin with, we know that the area of $\triangle IAJ$ is the same as the area of $\triangle IGH$, which is half the area of the square *IFGH*.

And assuming the line *L* meets the side *AB* at *K*, we can say that the area of $\triangle IAJ$ is half the area of the rectangle *AKJI*.

And next, we know that the area of $\triangle BCJ$ is the same as the area of $\triangle FCD$, which is half the area of the square *CDEF*.

And since the line *L* meets the side *AB* at *K*, we can say that the area of $\triangle BCJ$ is half the area of the rectangle *KBCJ*.

So we can say that the area of the square *ABCI* is the same as the sum of the areas of the two squares *CDEF* and *IFGH*.

And of course, there can be many other ways we can show the formula can hold.

And in fact, the former US President, James A. Garfield (the 20th) showed that it can hold, too. He used the idea of the area of a trapizoid.

Fig. 6.B

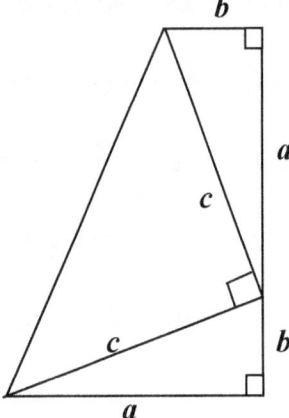

The trapozoid above is made of three right triangles. One is c by c, and the other two are a by b each. So the area of the trapizoid is the same as the sum of the areas of the three right triangles.

To begin with, the area of the trapizoid is half the product of the height and the sum of the two sides parallel to each other.

And the hegith is the distance between the two sides paralllel to each other. Of the two sides parallel, the lower is often called the base, and the upper is called the top.

So in sum, the area of a trapizoid is half the product of the height and the sum of the base and the top.

And thus, the area of the trapizoid above is: $\dfrac{(a+b)(a+b)}{2}$, which is: $\dfrac{(a+b)^2}{2}$, because the sum of the two sides parallel is: $a + b$, and the height is $a + b$, too.

Next, the sum of the areas of the three right triangles is: $\dfrac{c^2}{2} + \dfrac{ab}{2} + \dfrac{ab}{2} = \dfrac{c^2}{2} + ab$.

And thus, we need to have: $\dfrac{(a+b)^2}{2} = \dfrac{c^2}{2} + ab$.

So we get : $(a + b)^2 = c^2 + 2ab$. And we have: $(a + b)^2 = a^2 + 2ab + b^2$.

Thus, we get : $a^2 + 2ab + b^2 = c^2 + 2ab \Rightarrow a^2 + b^2 = c^2$, which is the formula.

And in the figure below, we can see the reason that the aera of a trapizoid is half the product of the height and the sum of the base and the top. Putting together two identical trapizoids as the one above, we get a square, which is *a + b* by *a + b*, and is below:

Fig. 6.C

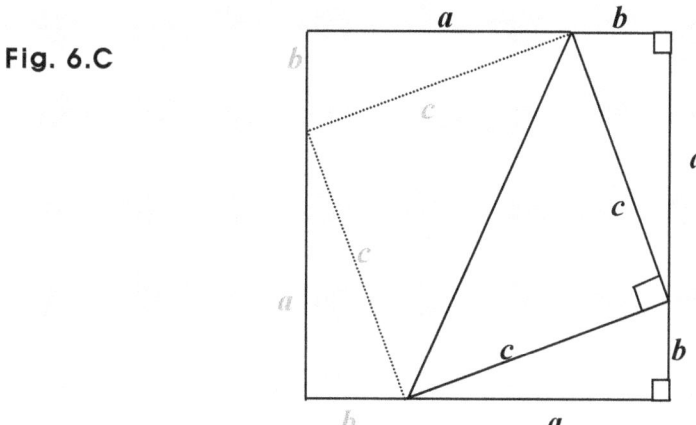

And the area of the square above is: *a + b* by *a + b*, and thus, is: $(a + b)^2$.

So the area of the trapizoid is half the area above, and thus, is: $(a + b)^2/2$.

And we can notice that we can make two identiacal quadrangles applying one cut to the trapizoid above. Cutting along a line perpendicular to the side on the left in the trapizoid below and passing trough *M*, we get:

Fig. 6.D

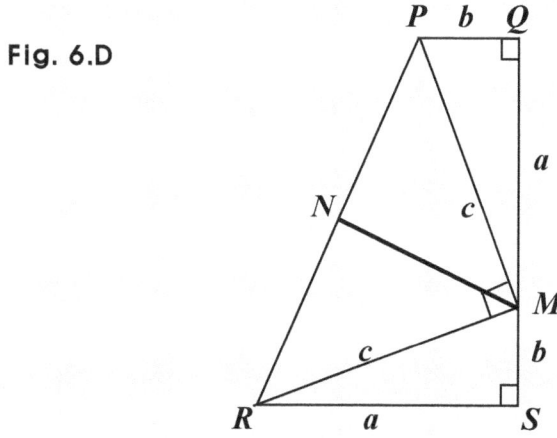

Then, we can see that the quadrangle ***PQMN*** is identical to the quadrangle ***MSRN***.

7. What is a circle?

A circle is a collection of all points that have equal distances from a particular point, the equal distance is called the radius of the circle, and the particular point is called the center of the circle.

And showing a circle in math, we usually put it in a coordinate plane as the *x-y* plane.

So for instance, collecting all points that are 3 cm away from a point (1, 2) in the *x-y* plane, we get a circle as below, so its radius is 3 cm, and its center is (1, 2).

Fig. 7.0

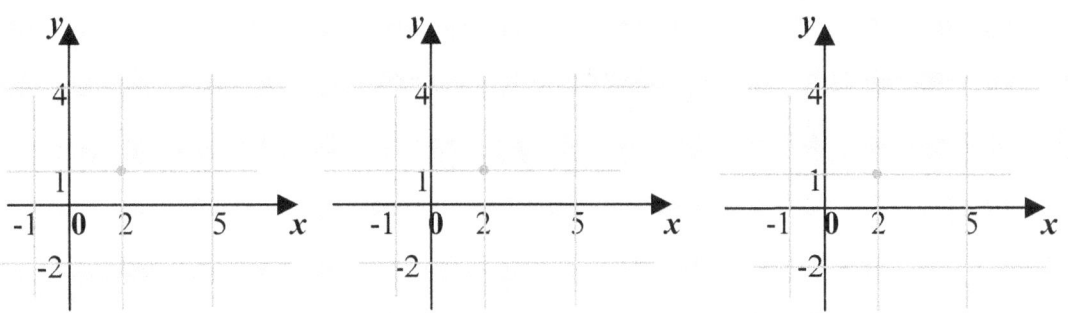

And as in the case of a triangle, we want to note that a circle is an idea and not a material object, and that saying just a circle, we mean a closed line segment and not a circular disk, which is full of points, and thus, is a plate. So *nothing* inside a circle is a part of the circle, which is therefore, a closed line segment only, and is empty inside.

Fig. 7.1

a circular disk

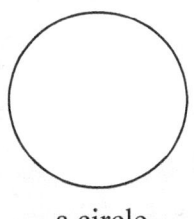

a circle

So a circle is a closed line segment, which is a collection of all the points, the distance from each of which to a particular point is the same. The distance is the radius, and the particular point is the center.

And we can take a circle for a polygon that is a collection of all the same and smallest line segments that are the same distance away from a particular point called the center of the polygon.

Technically therefore, circle can be taken as a regular polygon with infinitely many sides equal. So in short, a circle can be taken for an infinite regular polygon.

Fig. 7.2

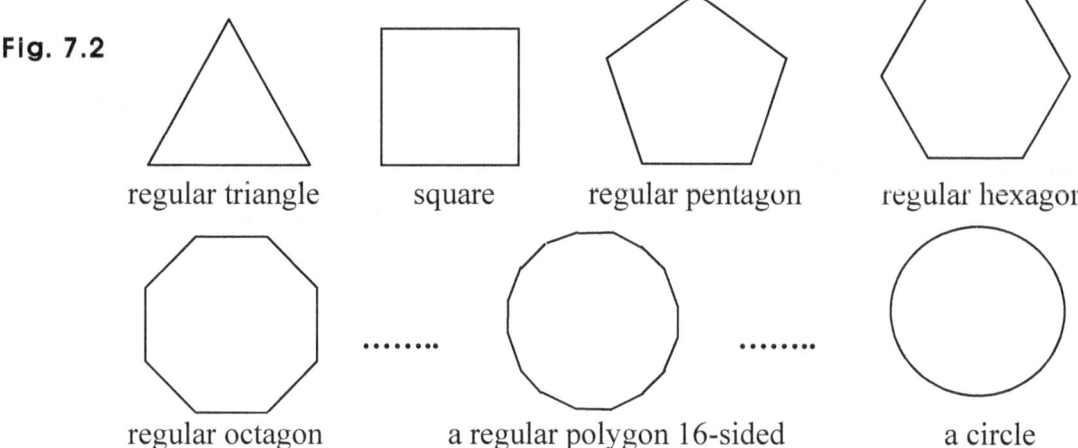

| regular triangle | square | regular pentagon | regular hexagon |

| regular octagon | a regular polygon 16-sided | a circle |

And we can put the diagram above the way below, too.

Fig. 7.3

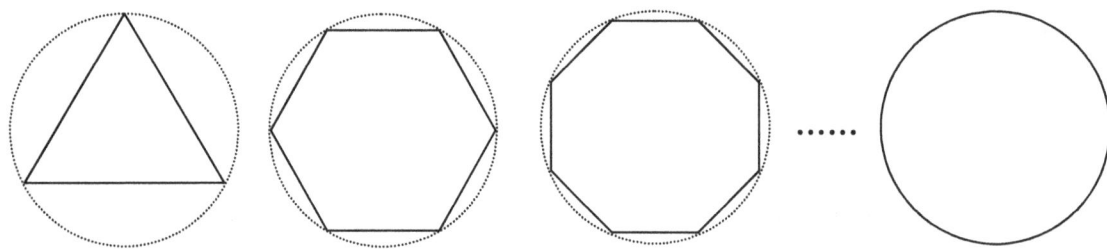

So a circle is a 2-D object, which is a collection of all points that are the same distance away from a particular point. And the particular point is called the center of the circle, and the same distance is called the radius of the circle. Also, a circle is said to have its diameter, which is a line segment passing through the center and connecting two points facing each other in the circle. So the diameter is twice the radius of the circle.

• What then, about the circumference of a circle?

It is no other than the circle. And more specifically, the circumference of a circle is the length of the curved and closed line segment forming the circle.

So for instance, assuming the radius is r, and C is the circumference, we get: $C = 2\pi r$, where π is the circular ratio, which is an irrational number, and is 3.141592… And the diameter is twice the radius. So in short, the circumference of a circle is: π times the diameter.

And saying just the radius of a circle, we mean the distance from the center to a point in the circle. So for instance, if the distance is 2, the radius is 2.

Saying however, a radius of a circle, we mean a line segment connecting the center and a point in the circle. How many points are there in a circle though?

Infinitely many, of course. So a circle can be said to have infinitely many radii, each of which has the same length. And the same is true for the diameter, too.

So saying just the diameter of a circle, we mean the length of a line segment that is between two points facing each other in the circle, and is passing though the center.

• And we can put the diameter of a circle this way, too:

The diameter is *the largest distance* between two points in the circle.

Saying however, a diameter of a circle, we just mean such a line segment as stated above, that is, we mean a line segment connecting two points facing each other in a circle.

So a diameter passes through the center of the circle, and every diameter in a circle has the same length, which is twice the radius.

And a circle has infinitely many of such line segments with the same length. And thus, a circle can be said to have infinitely many diameters having the same length.

• So *what can* be said to *define* a particular circle?

The radius and the center can define a particular circle. So for instance, we can have only one circle of radius 3 centered at **(1, 2)** in the *x-y* plane. And also, defining that circle, we can use an equation as follows: $(x - 1)^2 + (y - 2)^2 = 3^2$, which is therefore, called the equation of the circle of radius 3 centered at **(1, 2)** in the *x-y* plane. And the equation is in fact, the distance formula, and is said to be in the standard form.

How come is it though, the distance formula?

Assuming for instance, *a* and *b* are two legs in a right triangle, and *c* is the hypotenuse, we can put the distance formula this way: $a^2 + b^2 = c^2$.

So in the equation of the circle, $(x - 1)^2 + (y - 2)^2 = 3^2$, taking $(x - 1)$ as *a*, $(y - 2)$ as *b*, and 3 as *c*, we get: $a^2 + b^2 = c^2$.

And thus, $|x - 1|$ and $|y - 2|$ are the two legs in a right triangle where 3 is the hypotenuse, and 3 is the distance between an arbitrary point (x, y) to a point $(1, 2)$.

So the equation above indicates a geometric object, which is collection of all points, the distance from each of which to a point $(1, 2)$ is 3.

That is to say that (x, y) is an arbitrary point in the circle where the radius is 3, and the center of the circle is $(1, 2)$.

So using the distance formula, we can easily come up with the equation of any circle if given the radius and the center.

- What then, about the general form?

Putting it in the general form, we just expand or simplify the standard equation, and then, set the right hand side to 0. Thus, expanding or simplifying $(x - 1)^2 + (y - 2)^2 = 3^2$, and setting the right hand side to 0, we get: $x^2 + y^2 - 2x - 4y - 4 = 0$.

And we know that the radius of a circle is the distance from its center to a point in the circle. Why do we call though, the distance the radius?

That's probably because a point called the center of a circle seems to radiate terminal rays of the same length in all directions, and thus, the length is called the radius.

Fig. 7.4

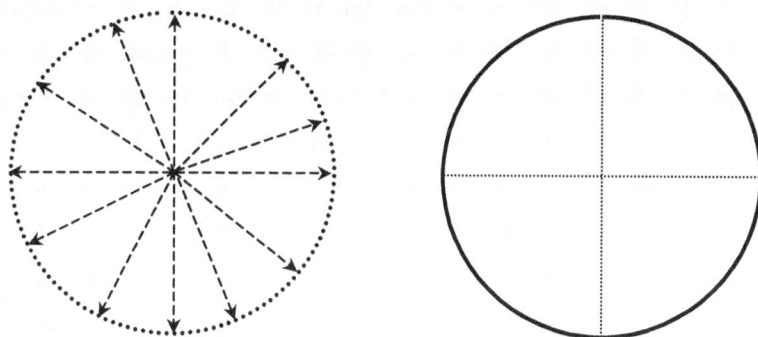

And each of the terminal rays is called a radius.
So the center of a circle can be said to radiate the same radii in all directions.

• And we can say that circles are in two kinds in connection with polygons: circumcircles and incircles.

A circumcircle surrounds a polygon, passes through all the vertices of the polygon, and is said to circumscribe the polygon, which is then, said to be inscribed in its circumcircle.

Some polygons have a circumcircle. So not every polygon has a circumcircle. However, every regular (equilateral) polygon has a circumcircle.

And many polygons can have the same circumcircle. In other words, many different polygons can share one particular circumcircle.

On the other hand, if a circle is tangent to every side of a polygon, and is inside the polygon, the circle is said to be inscribed in the polygon, and is called the incircle of the polygon.
And if a polygon has an incircle, the polygon is said to circumscribe the incircle, and is called a circumscribed polygon.

Some polygons can have an incircle. So not every polygon has an incircle. However, every regular (or equilateral) polygon has an incircle.

And of course, many polygons can have the same incircle at the same time. In other words, many different polygons can share one particular incircle, which is inscribed in all the polygons.

Fig. 7.5 Examples of regular polygons with the incircles and circumcircles

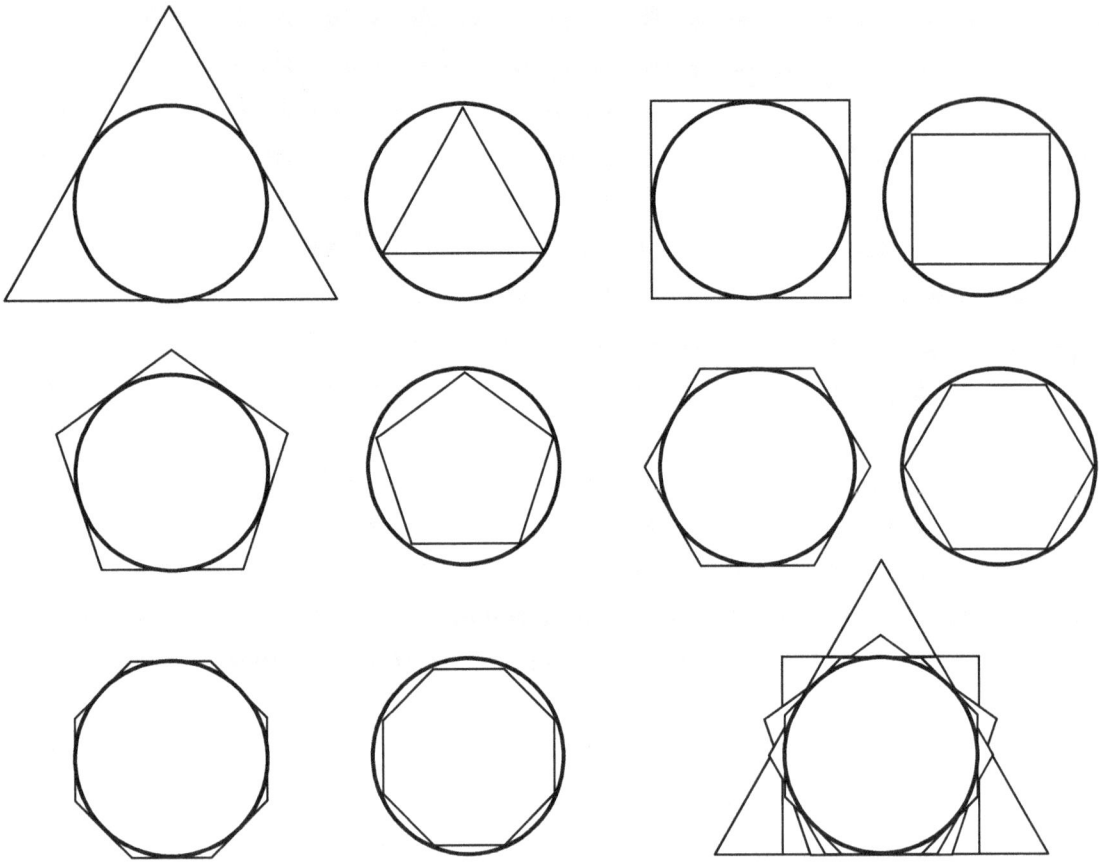

Fig. 7.6 Some circles can have a circumcircle only.

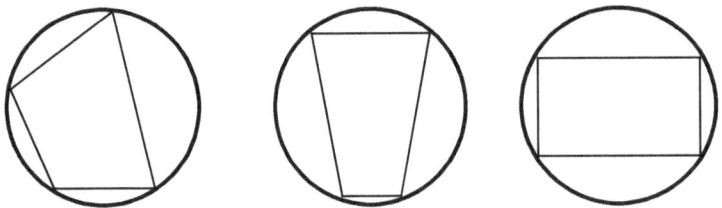

Fig. 7.7 Some polygons can have a circumcircle and an incircle both.

And in fact, if a circle is inscribed in a polygon, the polygon has its circumcircle.
And every triangle has an incircle and a circumcircle both. Some polygons however,
have *neither* an incircle *nor* a circumcircle.

Fig. 7.8

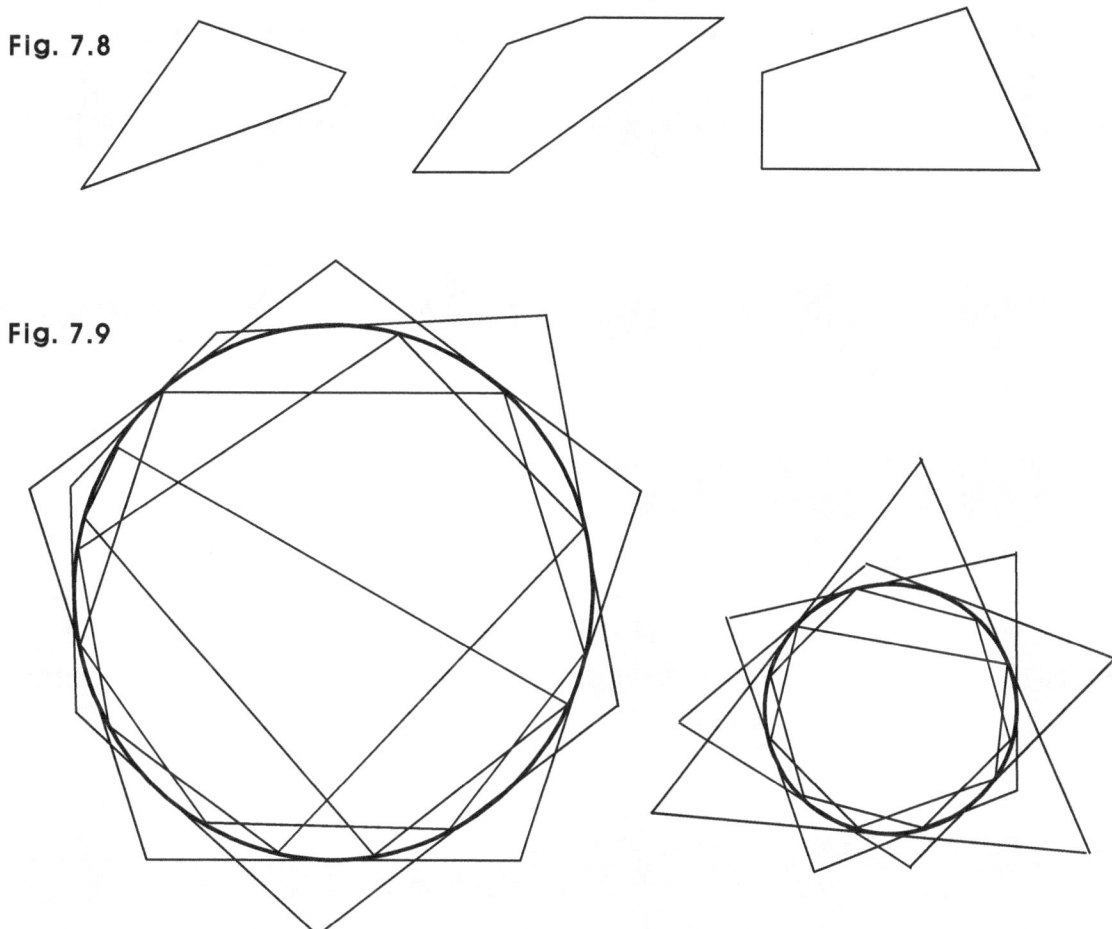

Fig. 7.9

A circle can circumscribe or can be inscribed in infinitely many polygons.
What is a polygon though?

A polygon is a set of there or more line segments forming a closed plane object.
We have three kinds in polygons. One is convex, another is concave, and the other is
complex.
Normally, just saying a polygon, we mean a polygon convex where every line segment
meets two of the others at its both ends, and every internal angle is between 0 and 180°.

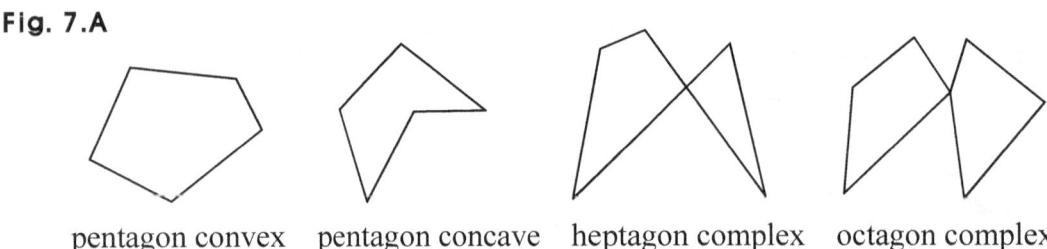

58

So for instance, every internal angle in a rectangle is 90°, and every internal angle in a regular triangle is 60°.

And thus, a polygon convex is a closed plane object, and is a set of three or more line segments, each of which meets two of the others at its both ends only, and is a plane object where every internal angle is between 0 and 180°. And a line passing through such a polygon can cross up to two sides in the polygon, so it cannot cross three or more sides.

Fig. 7.A

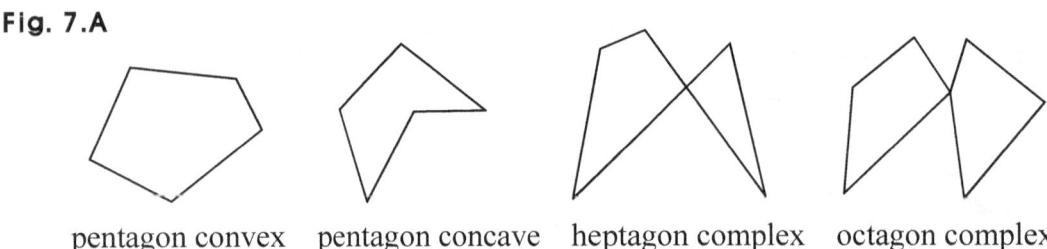

pentagon convex   pentagon concave   heptagon complex   octagon complex

• Now, a circle is a line segment closed, and is curved evenly so that every point in the circle is the same distance away from a point called the center. And the same distance is called the radius. So anyway, a circle is a line segment. How then, can we find the length of such a line segment?

We call such a length the circumference of a circle.
And the circumference is approximately 3 times the diameter of the circle.
So the ratio of the circumference to the diameter is approximately 3. And thus, assuming C is the circumference, and D is the diameter, we can set: $C \approx 3D$.
What then, is the exact value of the ratio?

The ratio is not in fact, a rational number as 3.14, and is an irrational number. And more precisely, the ratio is 3.141592…, and is often taken as 3.14 as an approximation.
And the ratio is called the circular ratio, and we often use a Greek letter π to indicate the ratio. So we often set: $\pi = 3.14$.

And thus, assuming C is the circumference, and D is the diameter, we get: $C = \pi D$.

And we know the radius is half the diameter. So assuming *r* is the radius, we can put the circumference this way, too: $C = 2\pi r$.

So if for instance, the radius of a circle is 1, the circumference is 2π.

And next, we know a circle is a closed line segment. So we should be able to find the area taken by a circle, too. How then, can we get the area?

We can partition a circular disk of radius *r* into circular wedges the way below:

Fig. 7.B

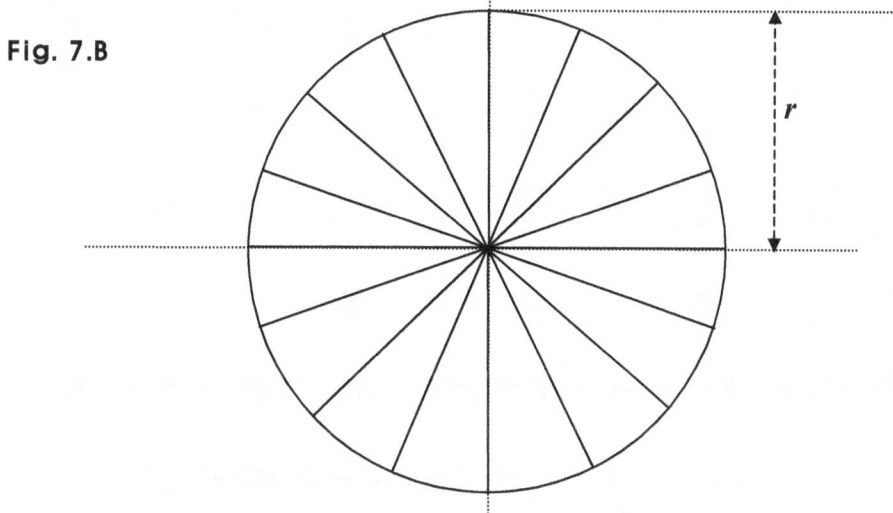

Then, assuming we partition the disk above into infinitely many wedges, we can take the wedges as triangles isosceles, and thus, can put together all the triangles the way below:

Fig. 7.C

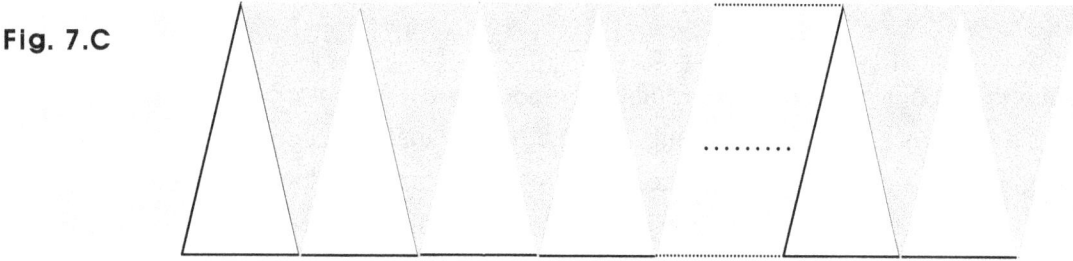

Then, we can say that all the triangles in gray are from the upper half of the disk, and that all the triangles in white are from the lower half of the disk.
And also, we can say that putting together all the triangles in gray and white, we get a parallelogram disk, which is however, no other than a rectangular disk.
What then, is the length of the base of the rectangle above?

We know that the circumference of the disk is $2\pi r$, and all the triangles in white are from the lower half of the disk. So the base of the rectangle is the half the circumference, and thus, is πr.   What then, is the height of the rectangle?

We know that we partition the disk above into infinitely many wedges. So we can take the wedges as triangles isosceles, and also, the height of the triangle can be taken as the radius of the disk, that is, r.   What then, is the area of the rectangle?

The area of a rectangle is the product of the base and the height.
And the base is πr, and the height is r. So the area is: πr^2.
And we know that circular disk is no other than the rectangular disk. So the area of the circle is no other than the area of the rectangle, and thus, is: πr^2.

And thus, assuming A is the area of a circle of radius r, we get: $A = \pi r^2$.

And also, assuming C is the circumference of a circle of radius r, we get: $C = 2\pi r$.

So what matters in the area or the circumference of a circle is the radius.
And thus, if two circles share the same radius, that is, have the same radii, can we say that the two circles are the same?

In analytic geometry (called coordinate geometry, too), we put a circle in a coordinate plane as the *x-y* plane. And working with a circle in analytic geometry, we can put it in an equation, too, which is called the equation of the circle, and is expressed in terms of *x* and *y* if we put the circle in the *x-y* plane. And defining a particular circle in analytic geometry, we specify not only the radius but the center, too.

So what matters in a circle in analytic geometry is not only the radius but the center, too.

And thus, even if circles have the same radii, that is, share the same radius, we say that the circles are different if the centers are different.

In other words, though the circles themselves are the same, if they have different centers, they are said to be different. That's because their equations are different, since their centers are specified in their equations.

So for instance, putting a circle in an equation $(x - a)^2 + (y - b)^2 = r^2$, we mean a circle of radius r centered at a point (a, b).

So for instance, putting a circle in an equation $(x - 1)^2 + (y + 1)^2 = 2^2$, we mean a circle of radius 2 centered at a point $(1, -1)$.

And the same is true for many other curves, too. So even if the curves themselves are the same, if their equations are different, we say that the curves are different.

What if however, circles share the same center?

If circles have the same radii and the same centers, that is, if they share the same radius and the same center, we say that they are the same circles.

If however, the circles share the same center, but have different radii, the circles are said to be concentric. That is, they are called concentric circles.

So concentric circles share the same center, but their radii are different.

Fig. 7.D

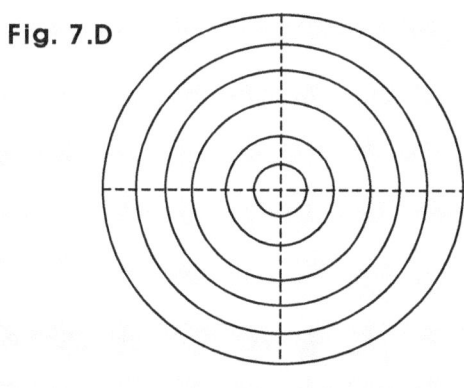

concentric circles

And the space (area) between concentric circles is called an annulus.

8. Circles and Angles

So to begin with, what is a circle?

A circle is a 2-D object, which is a collection of all points that are the same distance away from a particular point. And the particular point is called the center of the circle, and the same distance is called the radius of the circle. Also, a circle is said to have its diameter, which is a line segment passing through the center and connecting two points facing each other in the circle. So the diameter is twice the radius of the circle.

What then, about the circumference of a circle?

It is the length of the curved and closed line segment forming the circle. For instance, assuming the radius is r, and C is the circumference, we get: $C = 2\pi r$, where π is the circular ratio, which is an irrational number, and is $3.141592\ldots$

Why is though, the same distance stated above called the radius?

Saying *the* radius of a circle, we mean the distance from the center to a point in the circle.

That's because every distance from the center to every point in a circle is the same.

Saying however, just a radius of a circle or a radius in a circle, we mean a line segment connecting the center and a point in the circle. How many points are there in a circle?

Infinitely many, of course. So the center of a circle can be said to *radiate* infinitely many line segments, each of which has the same length. And thus, probably for that reason, we call each of the line segments a *radius*, and call the same length *the* radius.

Suppose now, all the radii in a circle are terminal rays, that is, rays with finite lengths, and all the rays begin at the center of the circle.

Then, all their lengths are equal, so all their terminal points are in the circle.

That is to say that all the terminal points form the circle. And we can say that the center emits such rays in all directions.

Fig. 8.0

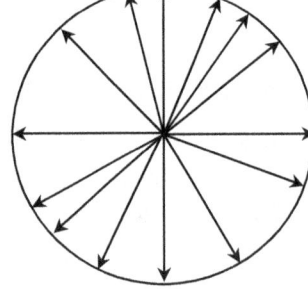

Also, we can say that each of all the rays has a different direction. What then, do we call the difference between the directions of two rays?

We call it an angle. More specifically, we call it the angle between the two rays. And thus, we can say that *a difference in direction is an angle*. And we know that the rays are the radii in the circle. So we can say that every radius in a circle has a different direction. What then, can we say a pair of radii can form?

Two radii in a circle can be said to form a geometric object, which can show a difference in direction. What geometric object though?

Two radii in a circle have different directions. And a difference in direction is an angle.

Fig. 8.1

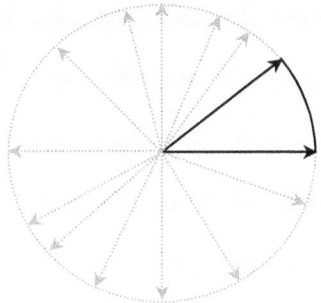

So as shown above, an angle can be indicated by a circular wedge made of two radii and a part of a circle. And thus, a pair of radii in a circle can be said to form a geometric object called an angle.

Is there anything else though, we can call an angle?

Suppose next, a terminal ray is turning about the origin in the *x-y* plane.

What then, do we call the amount of turning the ray makes?

We call it an angle, too. So we can say that an angle is an amount of turning, too.

Fig. 8.2

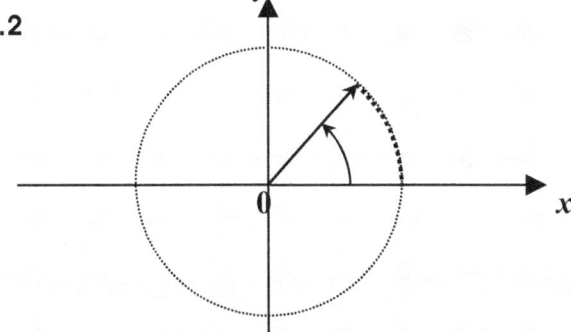

And also, we can say a difference in direction is an amount of change in direction. So in short, an angle can be called a change in direction, too.

And thus, we can use an angle specifying an amount of turning or a change in direction. In short, *an angle* is *an amount of turning* or *a change in direction*.

(And we know a pair of radii in a circle form an angle, which is a geometric object. So we can put an amount of angle this way, too: An amount of angle can tell us how much two radii in a circle are apart from each other. For instance, if in a circle, a radius is 30° away from another radius in the circle, the angle between the two radii is 30°.)

And of course, expressing an amount of angle, we use a unit of measure as in the cases of other objects as lengths, areas, volumes, weights, etc. For instance, we use kilograms or pounds for weights, liters or gallons for volumes, meters or miles for lengths, and square meters or acres for areas, etc. What then, is the unit of measure for angles?

We have two kinds in such a unit. One is *degree*, and the other is *radian*, called *rad* for short. And thus, we can use degrees or radians for angles. So for instance, an angle can be 25 degrees, 1 degree, 3 radians or 3 rad, or 1 radian or 1 rad.

And using degrees and putting in writing an amount of angle, we use a symbol, which is a small circle, and we put it at the upper right-hand corner of the number indicating the amount of angle, that is, we use it as a superscript as in 25°. What then, about radians?

We use no symbol for radians. So assuming for instance, A is known to be an angle in radian, we just put it in writing this way: $A = 2$, which means A is 2 radians.

Having to clarify though, an angle is in radian, we put 'rad' after the number. So for instance, if an angle B is 3 radians, we can put it in writing this way: $B = 3$ **rad**.

And thus, we have two metrologies for angles, that is, two systems of measurement for angles. One is radian system, and the other is degree system. And one system can get converted to the other. So two different amounts in angle can mean the same angle. For instance, we have: **180° = π rad** where π is the circular ratio, which is 3.141592… And briefly, we put it this say: **180° = π**. How can we calculate though, an amount of angle?

As explained above, an angle is a geometric object, which is a wedge formed by two radii in a circle and a part of a circle. What then, do we call such a part?

We call it an arc, which is a figure that can show not only a difference or a change in direction but turning, too.

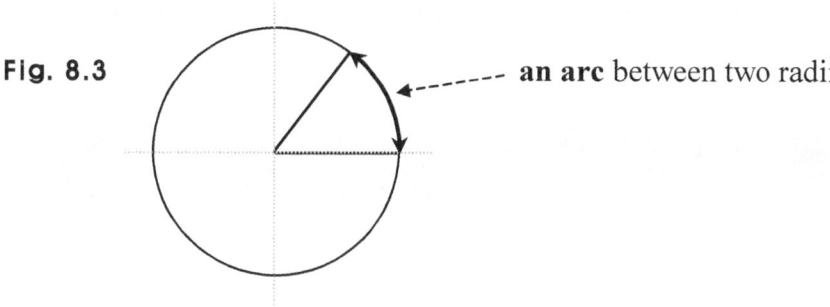

Fig. 8.3   **an arc** between two radii

And we can calculate an amount of angle by means of an arc and a circle.
More specifically, we can get an amount of angle using a ratio of an arc length to the circumference of the circle the arc belongs to. In short, an angle can be put in terms of a ratio of an arc to the circle that has the arc. How come though?

Let's get back to the ray turning about the origin in the *x-y* plane.

Fig. 8.4

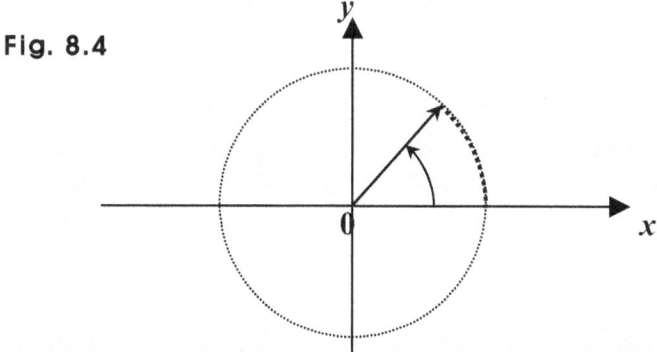

Then, first, an angle is an amount of turning. And an arc is said to have an angle. An amount of angle is however, not the length itself of an arc, because there can be many, infinitely many arcs between two radii in a circle. In the figure above, one of the two is the ray turning, and the other is on the *x*-axis.

And next, as the ray turns, the arrowhead makes an arc, and if the ray makes a complete turn, the arrowhead makes a circle.

So if the ray makes a complete turn, the length of the arc made is the circumference of the circle, that is, the arc is the circle itself. And if a half of a complete turn is made, the arc is half the circle. If a third of a complete turn is made, the arc is a third of the circle.

So an amount of turning, that is, the angle that an arc has is proportional to the ratio of the length of the arc to the circumference of the circle the arc belongs to.

That is to say that the angle an arc has is proportional to the ratio of the arc to the circle that has the arc. So in short, an angle is proportional to the ratio of an arc to a circle.

And thus, assuming θ is an angle, k is such a ratio, and C is a constant, we can set: $\theta = Ck$. What then, is the constant C?

It is $360°$. How come though?

If the ray makes a complete turn, that is, the ray gets for the first time, the same direction as the direction it had before turning, in other words, if the arc made is a circle, we say that the angle made is $360°$.

So in short, a complete turn is $360°$. And thus, if the ray has made two complete turns, that is, if the ray gets for the second time, the same direction as the direction it had before turning, the angle made is twice $360°$, that is, $720°$. What then, about a half of a complete turn?

If the ray gets for the first time, the direction opposite of the direction it had before turning, the ray makes a half of a complete turn, that is, the arc made is a half circle, so the angle made is a half of $360°$, that is, $180°$. So in short, a half turn is $180°$. And in turn, a quarter turn is: $360°/4 = 90°$. What then, about one and a half turn?

If the ray gets for the second time, the direction opposite of the direction it had before turning, it has to make another complete turn after the first half of a complete turn, so the angle made is: $180° + 360° = 0.5 \cdot 360° + 1 \cdot 360° = (0.5 + 1)360° = 1.5 \cdot 360°$.

Thus in short, one and a half turn is: $1.5 \cdot 360° = 1.5 \cdot 2 \cdot 180°$, which is: $3 \cdot 180° = 540°$.

What then, about a third of a complete turn?

The angle made is a third of $360°$, that is, $120°$. And thus, we can specify an angle by the product of $360°$ and the ratio of the arc to the circle the arc belongs to.

So for instance, assuming the length of an arc is 2, and 12 is the circumference of the circle where the arc belongs, we get: 2/12 of $360°$, which is 1/6 of $360°$, which is $60°$.

So assuming $\boldsymbol{\theta}$ is an angle, and $\boldsymbol{k}$ is such a ratio, we can set: $\boldsymbol{\theta = 360°k}$.

And we can put an angle this way, too:

An angle can be said to indicate how much two lines meeting at a point are apart from each other. If two lines are parallel to each other, the angle is 0. What if two line segments not parallel are away from each other, and thus, do not meet at a point?

A line segment is in a line, and lines not parallel meet at a point. So the angle between the two lines the two line segments belong to is the angle between the two line segments.

Fig. 8.5

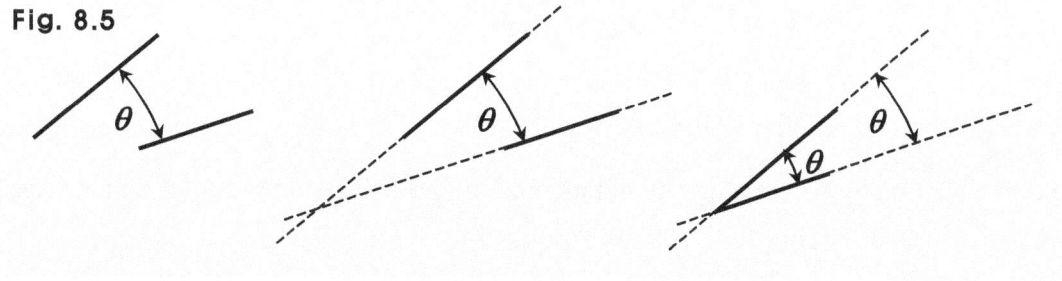

What then, about $1°$?

Dividing a circle into 360 equal parts around the center, we get 360 equal wedges. So taking one of the wedges, we get: $360°/360 = 1°$. So if the ray tuning counterclockwise makes a three hundred and sixtieth of a complete turn, we say that the angle made is $1°$.

And again, dividing one wedge worth $1°$ into 60 equal parts, we get 60 equal smaller wedges, each of which has an angle called 1 *minute* denoted by **1'**. So we get: $1°/60 = $ **1'**.

And thus, if the ray tuning counterclockwise makes a twenty thousand and six hundredth of a complete turn, we say that the angle made is 1'.

Then again, taking 1' apart into 60 equal parts, we get an angle called 1 *second* denoted by **1"**. So we get: $1°/360 = 1'/60 = $ **1"**.

And we can keep going taking the smaller wedges. We do not normally though, get a wedge worth less than 1 second, 1". What then, about $0.78°$?

It is an angle that is 78% of $1°$, so $0.78° = 78·1°/100$. And thus, we get: $0.1° = 1°/10$. So the bigger the wedge is, is the bigger the angle, in other words, the bigger the arc is, is the bigger the angle?

Not necessarily. Many different arcs can have the same angle.

Fig. 8.6

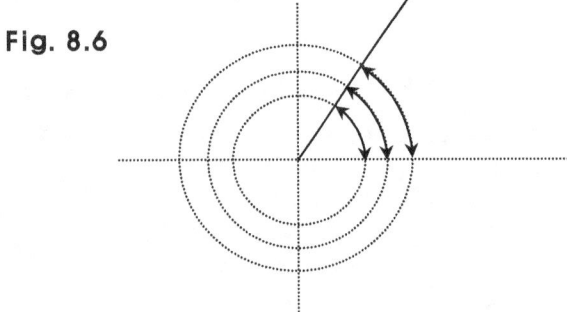

And thus, what matters in an angle is not an arc length itself but the ratio of an arc length to the circumference of the circle the arc belongs to.

That's because every circle has $360°$, but the radius of each can be different. So what?

First, many arcs with different lengths can have the same angle. All those arcs belong to circles said to be concentric. Concentric circles have the same center as shown above.

Now, holding an angle constant, we can see the larger the radius, the larger the arc.
That is, for a particular angle, as the radius grows, the arc grows, too.
How then, does the arc grows as the radius grows?

Getting back to the figure above, we can see three wedges, each of which is made of two radii and an arc. And extracting them, we get:

Fig. 8.7

Then, the three wedges are said to be similar to each other.
So the ratio between the radii is the same as the ratio between the arcs.

For instance, assuming R_1, R_2, and R_3 are the radii, and A_1, A_2, and A_3 are the arcs, and also, assuming $R_1 < R_2 < R_3$, and $A_1 < A_2 < A_3$, we get: $R_1 : R_2 : R_3 = A_1 : A_2 : A_3$.

And thus, we can get: $A_1/R_1 = A_2/R_2 = A_3/R_3$. What then, is the value of A_1/R_1?
In other words, if $A_1/R_1 = A_2/R_2 = A_3/R_3 = D$, what is D?

We know D is the ratio of each arc to its corresponding radius, and is constant.

And also, the angle in each of the three wedges, that is, the angle of each arc is the same, and thus, is constant. So we can reasonably expect that the ratio is the angle.

And the ratio is in fact, the angle. And thus, D is the angle.

We know however, a ratio is a number. How come an angle can be a number?

As mentioned earlier, we have two metrologies for angles, that is, two systems of measurement for angles. One is radian system, and the other is degree system.

And using the radian system, we can use as angles all real numbers, which are therefore, angles in radian. That was covered in the previous section. And we will see shortly how we get angles in radian.

Now, in such a circular wedge, the ratio of the arc to the radius is the angle of the arc. So assuming r is the radius, A is the arc, and θ is its angle in radian, we get: $A/r = \theta$.

That is, we get: $A = r\theta$.

And next, getting back to the concentric circles above, and holding the radius constant, that is, looking at one circle only, we can see the larger the arc, the larger its angle. More specifically, the arc is proportional to the angle.
And thus, assuming r is the radius, A is an arc, and θ is its angle, we get: $A = r\theta$.
So for instance, if the angle is 360°, we get: $\theta = 2\pi$, and thus, we get: $A = 2\pi r$.

Now, stating angles in terms of degrees, we put angles in degree system.
We have two systems where we can put angles. What then, is the other system?

It is called radian system, where we put angles in terms of radians. So 1 radian is a unit of measure for angles. What is 1 radian though?

Going back to one of the wedges stated above, we can get:

Fig. 8.8

Now, if the arc length is the radius of the circle that has the arc, the angle the arc has is 1 radian. That is, if an arc length is the radius, the angle of the arc is 1 radian.

So in short, if an arc is the radius, the angle is 1 rad.

And thus, if a circular wedge is like a regular (equilateral) triangle where three sides are equal, the angle of the arc is 1 rad. And we can take any angle based on the idea where we get 1 radian the way stated above: if an arc is the radius, we get 1 rad.

So it is probably the case where getting angles the way above, we call the angles radians.

That is, 1 radian is **radius angle**, the angle of the arc the length of which is the radius. What exactly is the way though, we get such angles in radian?

Such an angle is a ratio.

In radian system, an angle is the ratio of the arc to the radius in a circular wedge.

So we have: $A/r = \theta$ where A is the arc, r is the radius, and θ is the angle A has.

And thus, assuming $A = r$, we get: $\theta = A/r = r/r = 1$, which is called 1 radian. What angle in the degree system then, is 1 radian? That is, what degree is 1 rad?

Assuming first, the arc is a sixth of a circle of radius r, we get: $A = 2\pi r/6 = \pi r/3$, so we get: $\theta = A/r = (\pi r/3)/r = \pi/3$, called $\pi/3$ radian. And we know the angle the arc has is $60°$.

So we get: $60° = \pi/3$ rad. What then, about a quarter circle?

The length of a quarter circle of radius r is $2\pi r/4$, and the angle a quarter circle has is $90°$. So we get: $\theta = A/r = (2\pi r/4)/r = (\pi r/2)/r = \pi/2$, and thus, we get: $90° = \pi/2$.

And by the same token, since the angle of an eighth of a circle is $45°$, if the arc is an eighth of a circle, its angle θ is: $A/r = (2\pi r/8)/r = \pi/4$, and thus, we get: $45° = \pi/4$.

And also, the arc can be a half circle. Then, the circular wedge is a half of a circular disk.

So assuming $A = 2\pi r/2 = \pi r$, we get: $\boldsymbol{\theta} = A/r = \pi r/r = \boldsymbol{\pi}$, called π radian. And we know the length of a half circle of radius r is πr, and the angle a half circle has is $180°$. So we get: $180° = \pi$.

And of course, the arc itself can be a circle, too, that is, the wedge can be a circular disk. Then, we get: $A = 2\pi r$, so its angle $\boldsymbol{\theta}$ is: $A/r = 2\pi r/r = 2\boldsymbol{\pi}$, and thus, we get: $360° = 2\pi$ rad, since the angle a circle has is $360°$.

So next, what angle in degree is 1 rad?

We have: $60° = \pi/3$ rad, $90° = \pi/2$ rad, $45° = \pi/4$ rad, $180° = \pi$ rad, $360° = 2\pi$, etc. And using any of the ones above, we can get the angle in degree equivalent to 1 rad.

Using the fourth one above, we get; $180° = \pi$ rad $\Rightarrow$ 1 rad $= 180°/\pi = 57.29657...°$. So we get: 1 rad $\approx 57°$. What angle then, in radian is $1°$?

We have: $45° = \pi/4$ rad. So we get: $1° = \{(\pi/4)/45\}$ rad $= (\pi/180)$ rad $= 0.01745...$ rad. So we get: $1° \approx 0.0175$ rad. Why angles in degrees or radians though?

We can use either system working with angles. Working with angles though, we can say in large that we use angles in two different cases. In one case, we use angles that do not change, and in the other, we use angles that can change.

And angles that do not change can be said to be static, and angles that can change can be said to be dynamic. And using angles static, we are likely to use the degree system, and using angles dynamic, we are likely to use the radian system. In either case though, we can still use either system.

Using angles in degree, we have to use a symbol, which is a small circle, and need to put it upper right corner of the number indicating the amount of the angle as in $90°$.

Using angles in radian however, we just use real numbers as the angles. Usually, using angles in radian, we just use numbers only, and do not put rad or radian to the numbers.

So for instance, assuming A is π radian, we just put it this way: $A = \pi$, which represents a number 3.141592…, which is an irrational number, which is a real number. And for another instance, assuming B is 7 rad, we simply put it this way: $B = 7$. What then, can be the advantage of using angles in radian?

Suppose we work with a function, where the inputs are angles. Then, the input variable is an angle that can change, and thus, is dynamic (characterized by continuous change). And it's convenient to use real numbers as angles. So in such cases, we normally use angles in radian rather than those in degree.

Let's now get back again to the ray turning about the origin in the *x-y* plane, and look at angles from a bit different perspective.

Fig. 8.9

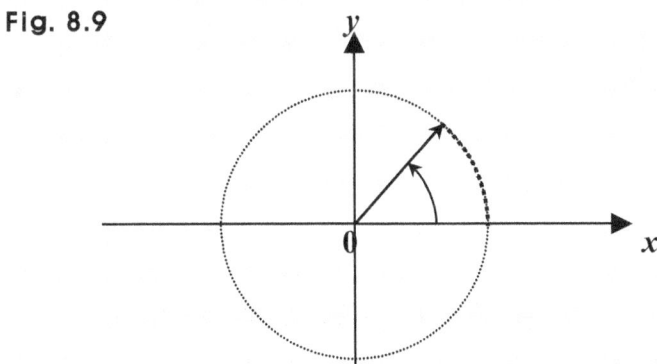

Then, we can see that the ray keeps changing its direction, and that the angle between the ray and the *x*-axis changes as the ray turns.

We know that the *x*-axis is fixed. So what makes an angle is the ray turning, and thus, we can say for simplicity that the angle between the ray and the *x*-axis is the angle of the ray, which is the angle the ray makes.

And initially, that is, when the ray is resting on the *x*-axis, its angle is assumed to be 0°. Suppose now, the ray starts turning counterclockwise.

Then, as the ray keeps turning, it keeps changing its direction, and its angle keeps changing. And in this case (tuning counterclockwise), the angle is said to be positive and increasing. And knowing the angle at a particular moment, we can specify the direction of the ray, at the particular moment, of course.   How then, do we get the angle?

Technically, a circle is an arc, too.

And by definition, a circle has 360°, which is 2π rad, and a half circle has 180°, which is π in radian.

So if the ray tuning counterclockwise makes a complete turn (or rotation) about the origin, its arrowhead makes a circle, and thus, we say the angle made is 360°, that is, 2π. And if two complete turns are made, the arrowhead makes a circle twice, so the angle made is twice 360°, and thus, is 720°, that is, 4π. Why not though, 360° but 720°?

The angle made in this case is the amount of turning, which is in this case, twice a complete turn. So the angle can be more than 2π, and for instance, can be 3π, $7\pi/2$, etc.

What if the ray is turning clockwise?

Then, its angle is said to be negative. So for instance, if the ray makes a complete turn clockwise, its angle is -360°, that is, -2π rad, and if a quarter of a complete turn is made clockwise, the angle made is -90° or $-\pi/2$.

And thus, tuning the ray counterclockwise or clockwise, we can get all angles. And indicating all such angles, we can use all real numbers.   So what?

For instance, we can come up with functions where inputs are angles. And it's convenient to use numbers as inputs. So in such a function, using as inputs angles in radian, we can use all real numbers as such inputs.   What are angles for though?

Working with objects that have to do with directions or that can change their directions, we need to work with objects called angles. We don't just work with angles though, of course. The angle we get to work with in such a case is one of the angles in a triangle. It's not just any triangle though.   What triangle then?

Of all kinds in triangles, the simplest and the most basic, and thus, the most fundamental is a triangle where one angle is 90°, called a right angle.

So such a triangle is called a right triangle.

And a right triangle is the place where a special geometry called trigonometry begins.

In short:

To begin with, by definition, putting angles in degrees, we get 360° if the ray makes a complete turn.
If the ray does not make any tuning, that is, if it is just sitting on the *x*-axis, the angle made is 0°. If the ray is turning counterclockwise, the angle made is said to be positive, and if the ray is turning clockwise, the angle made is said to be negative. So for instance:

If the ray is turning clockwise, and the ray makes a twelfth of a complete turn, the angle θ made is: **$-360^o/12 = -30^o$**.

And next, angles in radians are ratios.   What ratios?

An angle in radian is a ratio of an arc to the radius of a circle that has the arc.
So assuming A is the arc, r is the radius, and θ is the angle A has, we get: **$\theta = A/r$**.

And such a ratio can be negative or 0 as well as positive.
If the ray does not make any tuning, that is, if it is just sitting on the *x*-axis, the angle made is 0. If the ray is turning counterclockwise, the angle made is said to be positive, and if the ray is turning clockwise, the angle made is said to be negative, and thus, we get: $A = r|\theta|$, because A is a length, which is the length of the arc made, of course.

Thus for instance:

If r is the length of the ray turning clockwise, and the ray makes a twelfth of a complete turn, the arc A made is: **$2\pi r/12 = \pi r/6$**, so the angle θ made is $-A/r = -(\pi r/6)/r = -\pi/6$.
And we know putting the angle in degree, we get: **-30°**. So we get: **$-30^o = -\pi/6$**.
And thus, we get: **$180^o = \pi$** radian.

And the arc A can be half the circle or bigger.   So:

If r is the length of the ray turning, and the ray makes a complete turn either clockwise or counterclockwise, the arc A made is: $r|\theta| = 2\pi r$, since the angle θ made is -2π or 2π.

And if r is the length of the ray turning, and the ray makes one and a half of a complete turn either direction, the arc A made is: $r|\theta| = 3\pi r$, since the angle θ made is -3π or 3π.

Examples 0

Put the angles below in radian.

0. 30°   1. 45°   2. 60°   3. 90°   4. 120°

5. 180°   6. 270°   7. 360°   8. 405°   9. 775°

Suggestions or Solutions
To the Problems in the Examples 0

To begin with, by definition, we have: $180° = \pi$ rad.

And normally, rad is omitted, so in short, we just set: $180° = \pi$.

And we know 180 degrees is 180 of 1 degrees, that is, $180°$ is 180 of $1°$s.

In other words, we have: $180 \cdot 1° = \pi$.

So next, putting $1°$ in radian, we get: $1° = \pi/180$.

And next, we can put an angle $A°$ this way: $A \cdot 1°$.

And we know: $1° = \pi/180$.

So we can get: $A° = A \cdot 1° = A(\pi/180) = (A/180)\pi$, which is in radian.

That is to say that we get: $A° = (A/180)\pi$.

And thus, we get:

0. $30° = (30/180)\pi = (1/6)\pi = \pi/6$.

1. $45° = (45/180)\pi = (1/4)\pi = \pi/4$.

2. $60° = (60/180)\pi = (1/3)\pi = \pi/3$.

3. $90° = (90/180)\pi = (1/2)\pi = \pi/2$.

4. $120° = 2 \cdot 60° = 2(\pi/3) = 2\pi/3$.

5. $180° = (180/180)\pi = \pi$.

6. $270° = 3 \cdot 90° = 3 \cdot \pi/2 = 3\pi/2$.

7. $360° = (360/180)\pi = 2\pi$.

8. $405° = 360° + 45° = 2\pi + \pi/4 = 9\pi/4$.

9. $775° = (775/180)\pi = (155/36)\pi = 155\pi/36$, which equals $4\pi + 11\pi/36$.

Examples 1

0. Assuming **XY** is a line segment, find the two angles *A* and *B*.

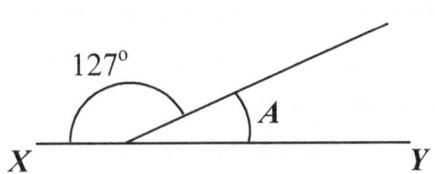 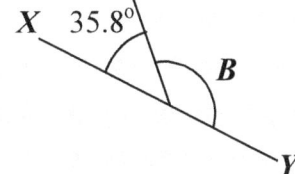

1. Assuming **XY** and **UV** are two line segments, show that the two angles *A* and *B* are the same

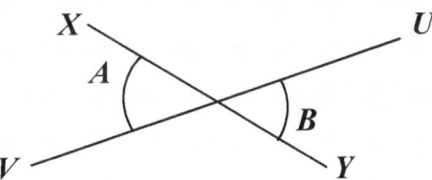

2. Assuming **XY** is parallel to **UV**, find the two angles *A* and *B*.

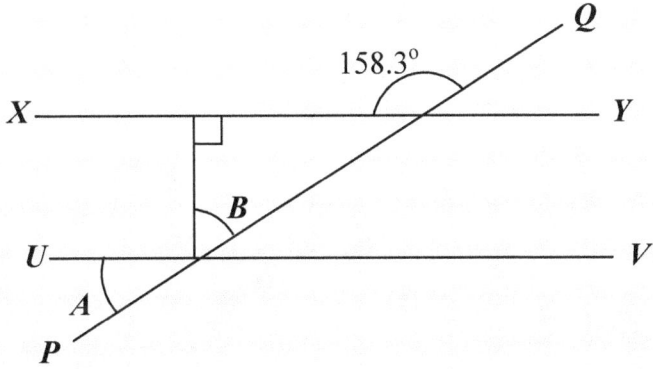

3. Assuming the line segment **AD** includes the side **AB** in **△ABC**, find ∠**CBD**.

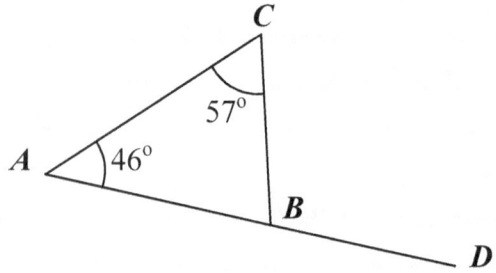

4. Assuming a rectangle gets folded the way below, find the angle **A**.

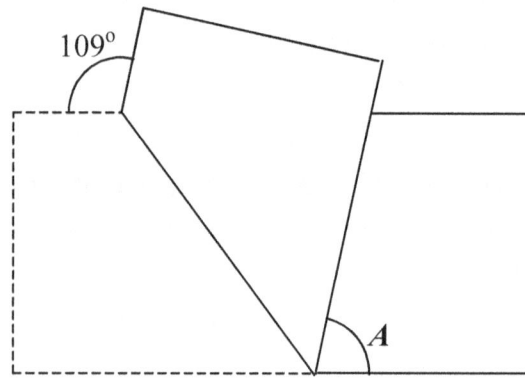

5. Assuming again, a rectangle gets folded the way below, and the angle **A** is the same as the angle **A** above, find the angle **B**.

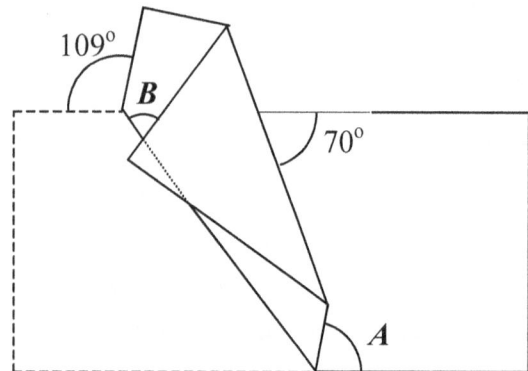

Suggestions or Solutions
To the Problem 0

Assuming *XY* is a line segment, find the two angles *A* and *B*.

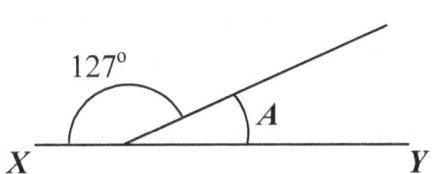 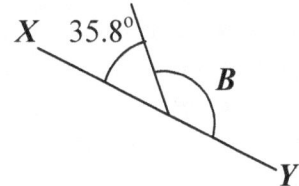

By definition, the angle between two line segments that are in the same line is $180°$. More specifically, the angle between two rays that are in the same line is $180°$ if their directions are opposite. If however, their directions are the same, the angle is $0°$.

Fig. 0.0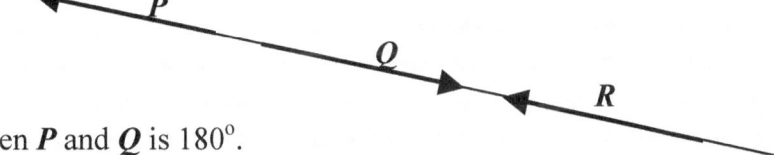

So the angle between *P* and *Q* is $180°$.
And the angle between *Q* and *R* is $180°$, too.
But the angle between *P* and *R* is $0°$.
Assuming however, *P*, *Q*, and *R* are *line segments*, we say that the angle between any two of the three is $180°$.

And thus, in the figure below, assuming *C* is a point in the line segment *XY*, we can say that the angle between *CX* and *CY* is $180°$.

Fig. 0.1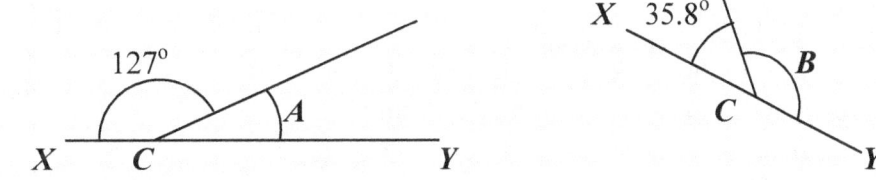

So we get: $127° + A = 180° \Rightarrow A = 53°$, and $35.8° + B = 180° \Rightarrow B = 144.2°$.

Suggestions or Solutions
To the Problem 1

Assuming *XY* and *UV* are two line segments, show that the two angles A and B are the same

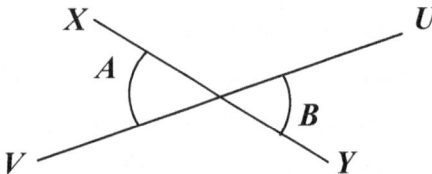

We know that *XY* and *UV* both are line segments.
So in the figure below, we get: $A + C = 180^\circ$, and $B + C = 180^\circ$.

Fig. 1.0

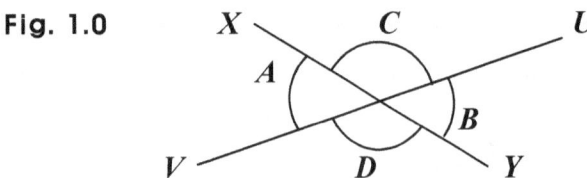

Thus, we get: $A + C = B + C \Rightarrow A = B$.
And by the same token, we can say that $C = D$, too.

And in particular, such two angles as *A* and *B* are called vertical angles, opposite angles, or vertically opposite angles. So vertical angles or opposite angles are the same.
Thus for instance, *C* and *D* are called vertical angles or opposite angles. And the angle *A* is said to be the vertical angle of the angle *B*. In short, *A* is vertical of *B*.

By the way, if two angles add up to 180°, the two angles are said to be supplement to each other. So for instance, the two angles *A* and *C* above are supplement to each other, and the angle *D* is supplement to the angle *A*. And the same is true, too, for the two angles *B* and *C*, and also, for the two angles *B* and *D*.

If however, two angles add up to 90°, the two angles are said to be complement to each other. So for instance, if $P + Q = 90^\circ$, *P* is complement to *Q*, and vice versa.

Suggestions or Solutions
To the Problem 2

Assuming *XY* is parallel to *UV*, find the two angles *A* and *B*.

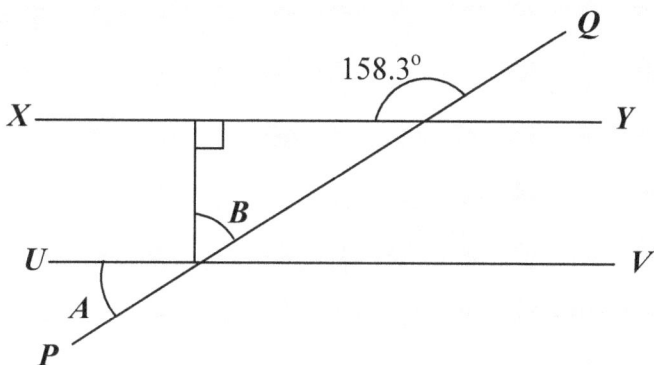

Indicating two angles *C* and *D* the way below, we can say that the angle *C* is supplement to the angle 158.3°, and the angle *D* is complement to the angle *B*, because *XY* is parallel to *UV*.

Fig. 2.0

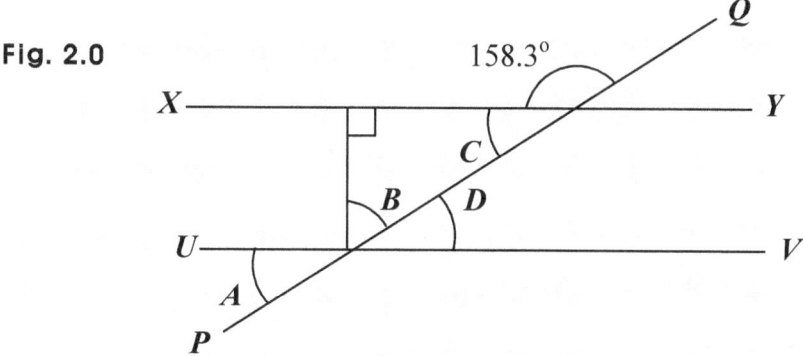

So we get: $C + 158.3° = 180°$, and $B + D = 90°$. And thus, we can get first: $C = 180° - 158.3° = 21.7°$. How then, can we get the angle *B*?

We know that the sum of the three angles in a triangle is 180°, and that *B* is one of the three angles in a right triangle, where one angle is 90°.

So we get:

$$B + C + 90° = 180° \Rightarrow B = 180° - C - 90° = 180° - 21.7° - 90° = 180° - 111.7° = 68.3°.$$

Next, we have: $B + D = 90^\circ$. So we get: $D = 90^\circ - B = 90^\circ - 68.3^\circ = 21.7^\circ$.

And next, we know that the angle A is vertical of the angle D. That is, $A = D$.
So we get: $A = 21.7^\circ$.

And we can notice that the angle A is the same as the angle C, which is not coincidental.
Both the two angles are equal, because the two lines XY and UV are parallel.
And we call such angles are corresponding angles, which are therefore, equal.

And also, we can notice that the angle C is the same as the angle D, which is not
coincidental either. That's also because the two lines XY and UV are parallel.
And we call such angles are alternate angles, which are therefore, equal, also.

So suppose for instance, that in the figure below, the three lines U, V, and W are parallel
to each other, and also, that the two lines S and T are parallel.

Fig. 2.1

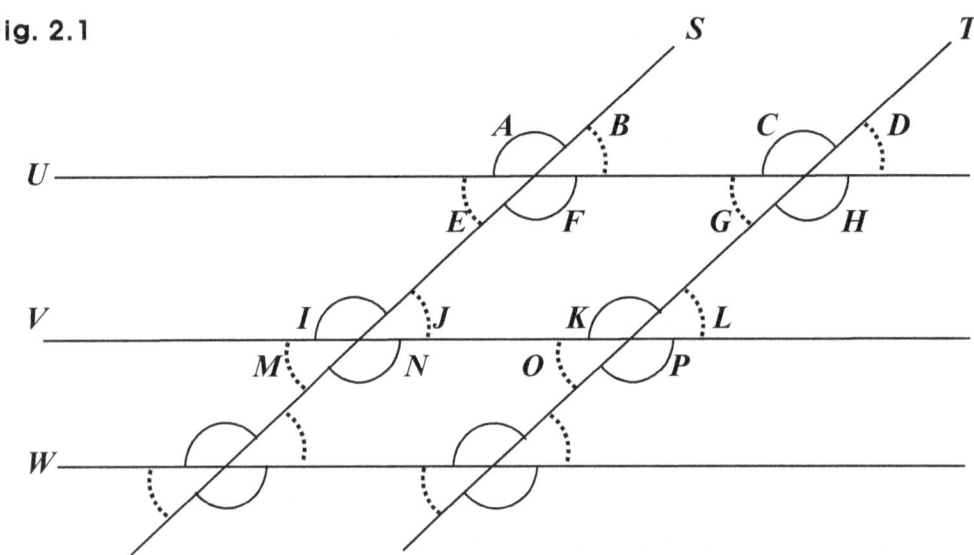

Then, we can say that A, I, C, and K are corresponding angles, E, M, G, and O are
corresponding angles, A and N are alternate angles, and E and J are alternate angles.

And we know that the two angles A and F are vertical angles, and thus, are equal.

So we get: $A = C = F = H = I = K = N = P$, and $B = D = E = G = J = L = M = O$.

Suggestions or Solutions
To the Problem 3

Assuming the line segment AD **includes the side** AB **in** $\triangle ABC$**, find** $\angle CBD$**.**

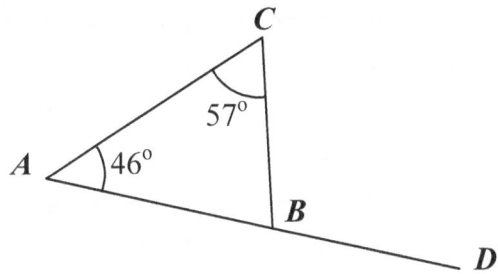

We know that the angle between AB and BD is 180°, because AB and BD are in a line.

So knowing $\angle ABC$, we can get $\angle CBD$, since $\angle ABC + \angle CBD = 180^{\circ}$.
How then, can we get $\angle ABC$?

We know that the sum of the three angles in a triangle is 180°.
So we get: $\mathbf{57^{\circ} + 46^{\circ} + \angle ABC = 180^{\circ} \Rightarrow \angle ABC = 180^{\circ} - 103^{\circ} = 77^{\circ}}$.

And thus, getting back to: $\angle ABC + \angle CBD = 180^{\circ}$, we get:

$$\angle CBD = 180^{\circ} - \angle ABC = 180^{\circ} - 77^{\circ} = 103^{\circ}.$$

And we can notice that $\angle CBD$ is the sum of the two angles 57° and 46°, that is, the sum of the two angles $\angle CAB$ and $\angle CBA$, which is not coincidental, though.

Assuming $X = \angle CAB + \angle BCA$, we get: $X + \angle ABC = 180^{\circ} = \angle CBD + \angle ABC$.

So we get: $X = \angle CBD \Rightarrow \angle CAB + \angle BCA = \angle CBD$.

And we know that $\angle ABC$ and $\angle CBD$ are supplement to each other, because we have:

$\angle ABC + \angle CBD = 180^{\circ}$. So we can say that $\angle CBD$ is supplement to $\angle ABC$.

And thus, we can say that in a triangle, the sum of two internal angles is the same as the external angle supplement to the other internal angle.

Fig. 3.0

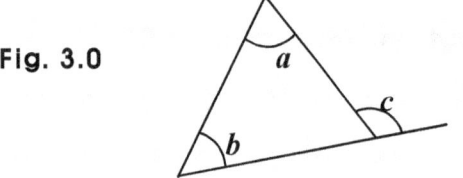

So in the figure above, we get: $a + b = c$.

Suggestions or Solutions
To the Problem 4

Assuming a rectangle gets folded the way below, find the angle _A_.

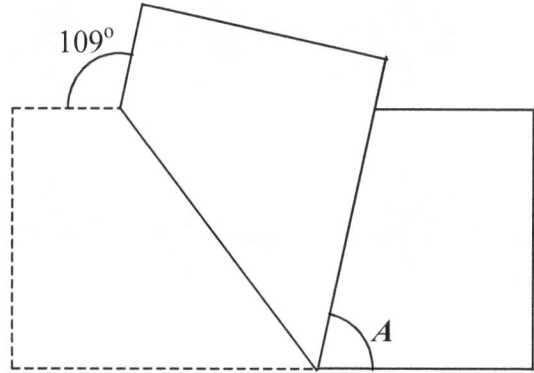

We know that the two sides facing each other in a rectangle are not only the same but parallel, too. So in the figure below, we have: _x // y_, and **_PQ // RS_**.

Fig. 4.0

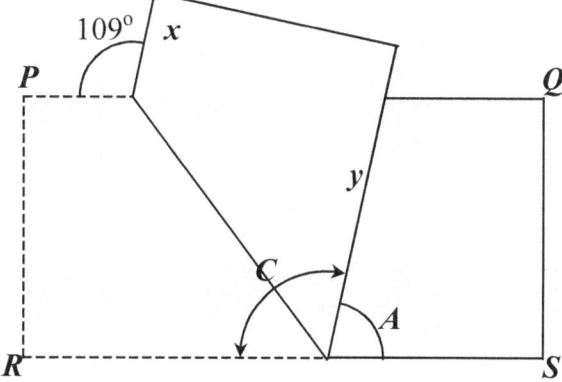

So we can say that the angle $109°$ and the angle _C_ are corresponding angles, and thus, are equal. That is, we get: **_C_ = 109°**.

And we know: **_C_ + _A_ = 180°**. So we get: _A_ = **180° − _C_ = 180° − 109° = 71°**.

Suggestions or Solutions
To the Problem 5

Assuming again, a rectangle gets folded the way below, and the angle *A* is the same as the angle *A* above, find the angle *B*.

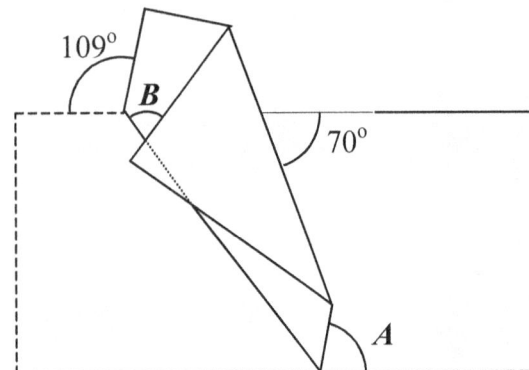

Unfolding the figure above, we can unfold the way to the solution.
So unfolding it the way below, we can see some angles that can help find the solution.

Fig. 5.0

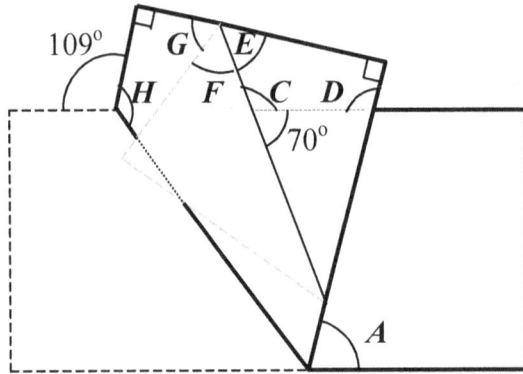

Then, to begin with, we get: $C + 70° = 180° \Rightarrow C = 110°$.

Next, we get: $D = 109°$, because D and $109°$ are corresponding angles, and thus, are equal. And we know that $C + D + E + 90° = 360°$ since the sum of all the four angles in a quadrangle is $360°$.

So we get: $E = 360° - 90° - C - D = 270° - 110° - 109° = 51°$.

Next, we get: $E = F$, because of unfolding. So we get: $F = 51°$, too.

Next, we have: $E + F + G = 180°$. So we get: $G = 180° - E - F = 180° - 102° = 78°$.

Next, we get: $B + G + 90° + H = 360°$. How come?

That's because B is one of the four angles in a quadrangle, and the sum of all the four angles in a quadrangle is $360°$. So we get: $B + G + 90° + H = 360°$.

In other words, we get: $B = 360° - G - 90° - H = 360° - 78° - 90° - H = 192° - H$.

So next, finding H, we can get B, which is the solution.
How can we get H though?

Unfolding the figure above, we can unfold the way to the solution.
So next, unfolding it the way below, we can see some angles that can help find the solution.

Fig. 5.1

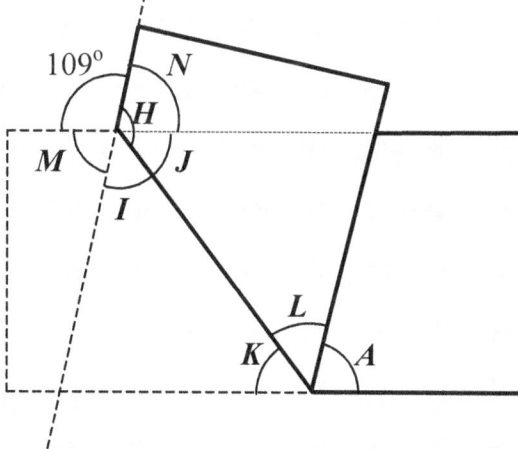

To begin with, we can see $K = L$ because after unfolding, L is the same as K.

Next, we know K and J are alternate angles, and so are I and L.
So we get: $K = J$, and $I = L$. Thus, we get: $I = J$, since $K = L$.

And we have: $I + J = 109^o$, because $I + J$ and 109° are vertical angles.

So we get: $2J = 109^o$, since $I = J$. Thus, we get: $J = 54.5^o$.

Next, we have: $M + 109^o = 180^o$. So we get: $M = 71^o$.

Next, we know: M and N are vertical angles. So we get: $M = N = 71^o$. What then, is the angle H?

It is the sum of N and J, so we get: $H = 71^o + 54.5^o = 125.5^o$.

Now, we have: $B = 192^o - H$. So we get: $B = 192^o - 125.5^o = 66.5^o$.

Examples 2

0.   Assuming two rectangles are placed the way below, find the angle *A*. If it cannot be found, explain why not.

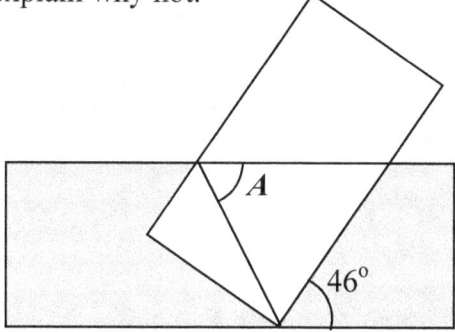

1.   Assuming two regular (equilateral) triangles and a regular pentagon are folded the way below, find the three angles *A*, *B*, and *C*.

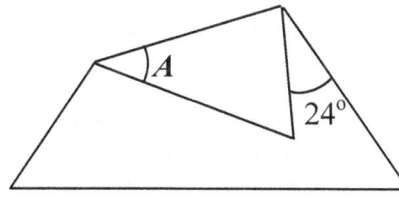

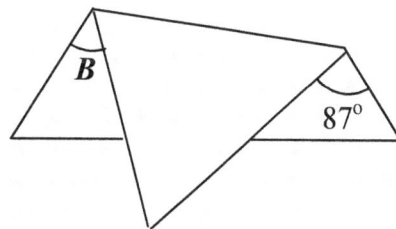

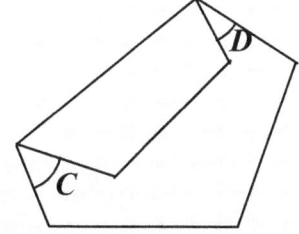

D = **23°**.

2. Assuming two regular (equilateral) triangles and a regular pentagon are folded the way below, find the three angles *A*, *B*, and *C*.

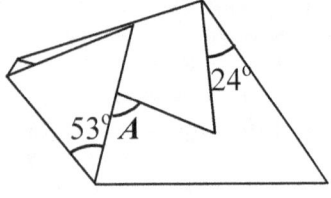

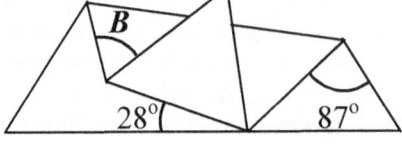

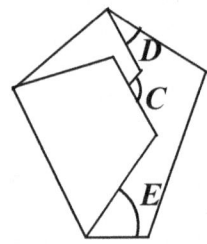

D = **23°**, and *E* = **45°**.

Suggestions or Solutions
To the Problem 0

Assuming two rectangles are placed the way below, find the angle A. If it cannot be found, explain why not.

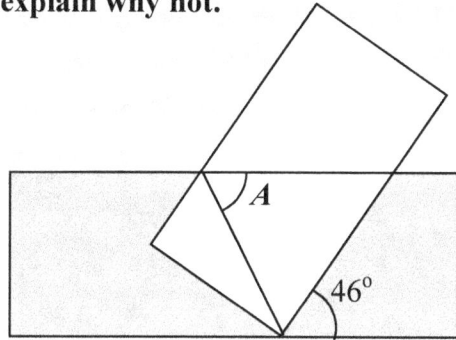

We cannot find the angle A. That's because the rectangles are not specific. In other words, the rectangles can have any dimensions, that is, they can have any lengths or widths. So the angle A cannot stay the same, and in turn, cannot have a particular value.

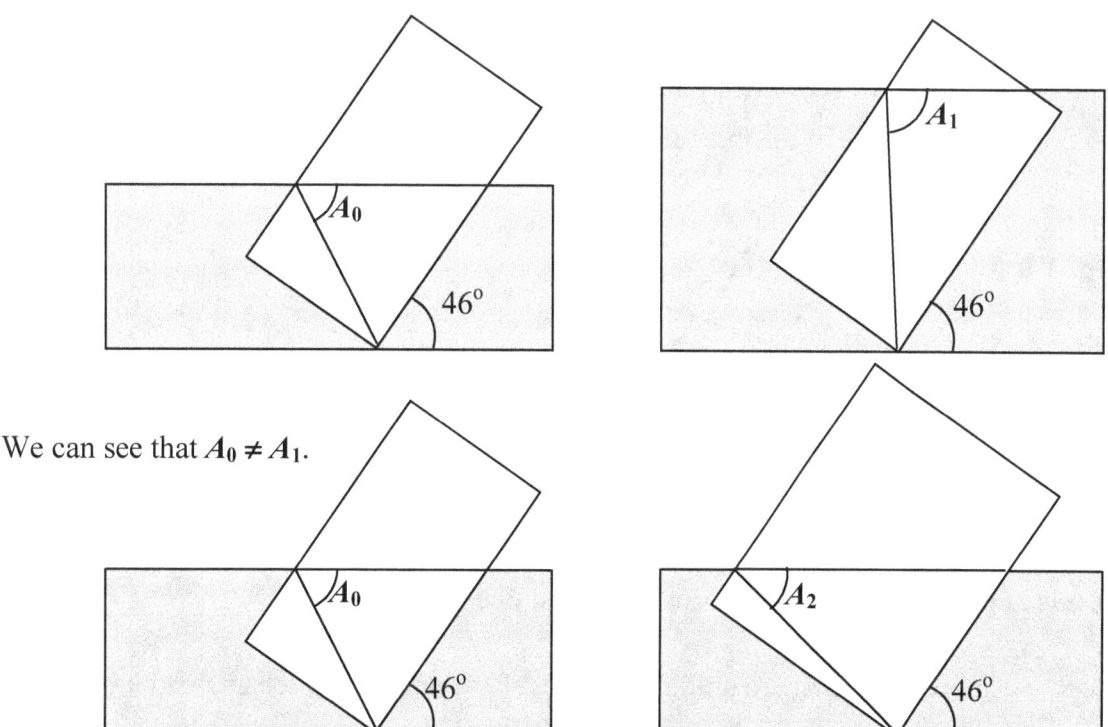

We can see that $A_0 \neq A_1$.

We can see that $A_0 \neq A_2$.

Suggestions or Solutions
To the Problem 1

Assuming two regular (equilateral) triangles and a regular pentagon are folded the way below, find the three angles *A*, *B*, and *C*.

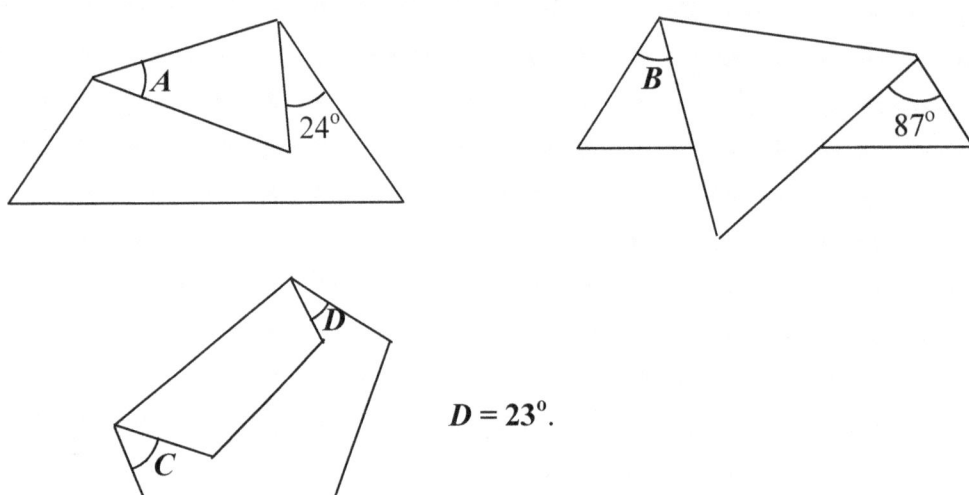

$D = 23^{\circ}$.

Unfolding the figures above, we can unfold the way to the solution.

So first, unfolding one of the three figures the way below, we can see some angles that can help find the angle *A*.

Fig. 1.0

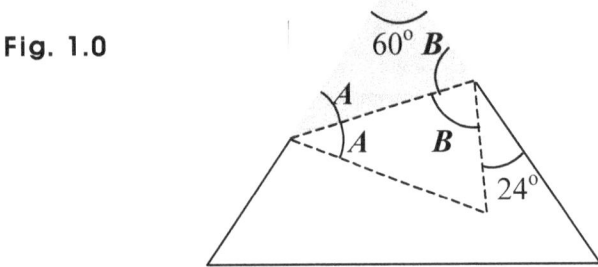

First, we can see two identical triangles.

One is in gray, and the other is made of dashed line segments.

Next, we know that the figure unfolded is a regular triangle. So each angle in it is 60°.

Next, we have: $B + B + 24^{\circ} = 180^{\circ}$. So we get: $2B = 180^{\circ} - 24^{\circ} = 156^{\circ} \Rightarrow B = 78^{\circ}$.

And next, we know that the three angles in a triangle add up to 180°.

So we get: $A + B + 60^\circ = 180^\circ \Rightarrow A = 180^\circ - 60^\circ - B = 120^\circ - 78^\circ = 42^\circ$.

And let's next, move on to the figure with the angle B, and unfold it the way below:

Fig. 1.1

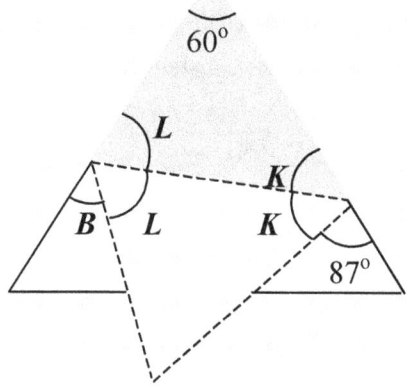

Then again, we can see two identical triangles.
One is in gray, and the other is made of dashed line segments.

And we know that the figure unfolded is a regular triangle. So each angle in it is 60°.

Next, we have: $K + K + 87^\circ = 180^\circ$.

So we get: $2K = 180^\circ - 87^\circ = 93^\circ \Rightarrow K = 46.5^\circ$.

And next, we know that the three angles in a triangle add up to 180°.

So we get: $L + K + 60^\circ = 180^\circ \Rightarrow L = 180^\circ - 60^\circ - K = 120^\circ - 46.5^\circ = 73.5^\circ$.

And next, we have: $L + L + B = 180^\circ$.

So we get: $B = 180^\circ - 2L = 180^\circ - 2 \cdot 73.5^\circ = 180^\circ - 147^\circ = 33^\circ$.

And let's next, move on to the figure with the angle C, and unfold it the way below:

Fig. 1.2

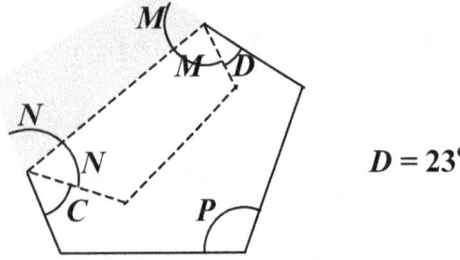

$D = 23^\circ.$

Then first, we can see two identical tetragons. One is in gray, and the other is made of dashed line segments. So we can indicate two pairs of same angles the way above.

Thus, we get: $M + M + D = 180^\circ$. And we have: $D = 23^\circ$, too.

So we get: $2M = 180^\circ - D = 180^\circ - 23^\circ = 157^\circ \Rightarrow M = 78.5^\circ$.

Next, we have: $N + N + C = 180^\circ$. So finding N, we can get C.
How can we find N though?

We know that the figure unfolded is a regular pentagon.
So every vertex angle in it is the same, and is P as shown in the figure above.
What then, is the angle P?

We can partition a regular pentagon into five triangles the way below:

Fig. 1.3

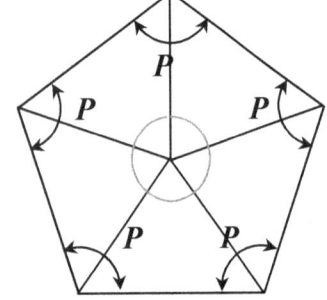

Then first, we assumed that in a regular pentagon, every vertex angle is P.

And next, we can see that at the center, the sum of all the five angles is 360°.
So taking the sum of all the angles in all the five triangles above, we get: $5P + 360^\circ$.

And also, we can put the sum this way, too: $5 \cdot 180^\circ = 900^\circ$. That's because the pentagon is made of five triangles, and the sum of three angles in each triangle is 180°.

So we get: $5P + 360^\circ = 900^\circ \Rightarrow P = (900^\circ - 360^\circ)/5 = 540^\circ/5 = 108^\circ$.

Now, we have: $N + N + C = 180^\circ$. So finding N, we can get C.
How then, can we find N?

Getting back to the pentagon we get unfolding the figure with the angle C, we have:

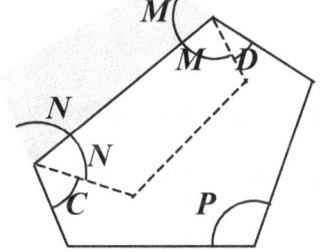

Fig. 1.4

Then, to begin with, the sum of all the angles in a tetragon is 360°, since we can make a tetragon putting two triangles together.
And next, every vertex angle is P, since the pentagon is regular.

$D = 23^\circ$.

So looking at the tetragon in gray above, we get: $N + M + 2P = 360^\circ$.

And we know: $M + 2P = 78.5^\circ + 2 \cdot 108^\circ = 294.5^\circ$.

So we get: $N = 360^\circ - (M + 2P) = 360^\circ - 294.5^\circ = 65.5^\circ$.

Now, we have: $N + N + C = 180^\circ$. So we get: $C = 180^\circ - 2N = 180^\circ - 131^\circ = 49^\circ$.

Suggestions or Solutions
To the Problem 2

Assuming two regular (equilateral) triangles and a regular pentagon are folded the way below, find the three angles *A*, *B*, and *C*.

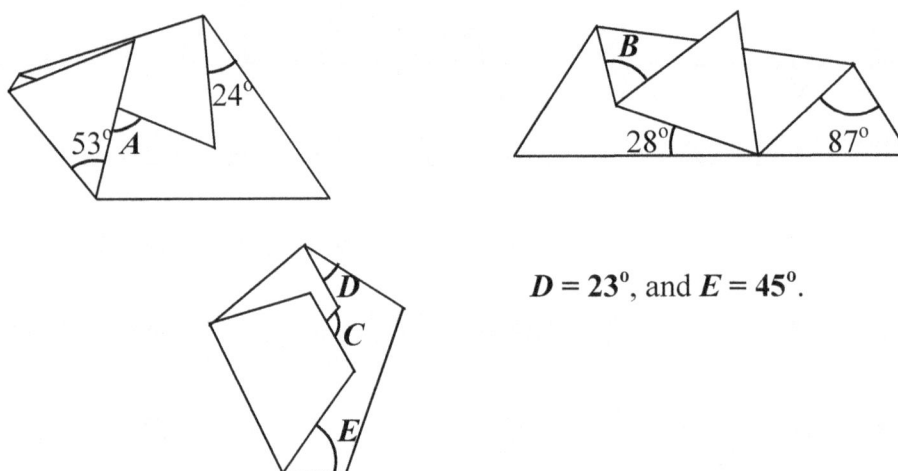

$D = 23^{0}$, and $E = 45^{0}$.

Unfolding the figures above, we can unfold the way to the solution.

So first, unfolding one of the three figures the way below, we can see some angles that can help find the angle *A*.

Fig. 2.0

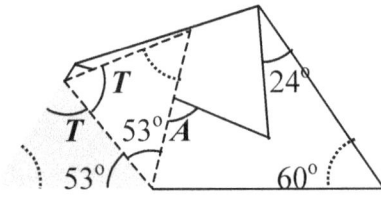

Then first, we can see two identical triangles.

One is in gray, and the other is made of dashed line segments.

Next, we know:

Unfolding the figure completely, we get a regular triangle.

Each angle in a regular triangle is 60^{0}.

And we know in any triangle, the three angles add up to 180^{0}.

So we get: $T + 53^{0} + 60^{0} = 180^{0} \Rightarrow T = 180^{0} - 113^{0} = 67^{0}$.

How then, can we get the angle *A*?

Let's consider the tetragon indicated by thick line segments below:

Fig. 2.1

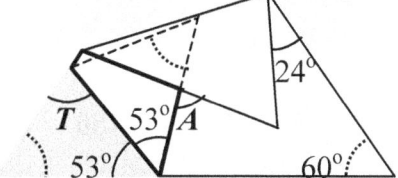

And zooming down on the tetragon, we can put it the way below:

Fig. 2.2

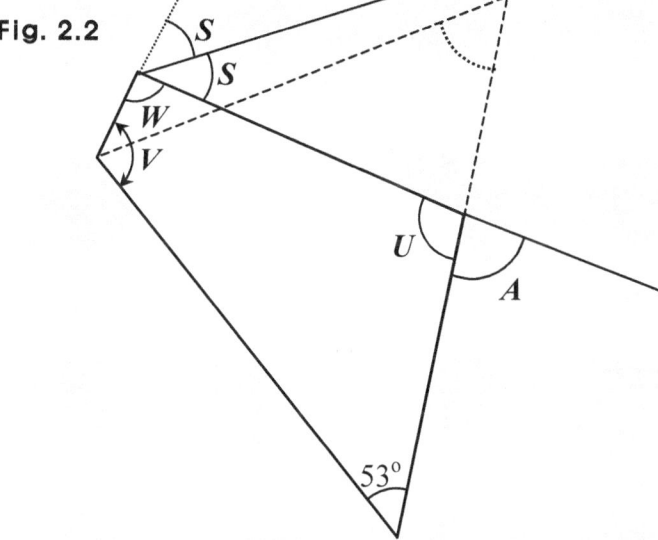

Then, assuming we can find the two angles *V* and *W* above, we can find the angle *U*. Then, we can get *A*, since *U* and *A* are supplement to each other, that is, $U + A = 180°$.

To begin with, the angle *S* is the angle *A* in the problem 1 above, and $A = 42°$. And we have: $S + S + W = 180°$. How come?

In the problem 1, we unfolded one of the three figures the way below:

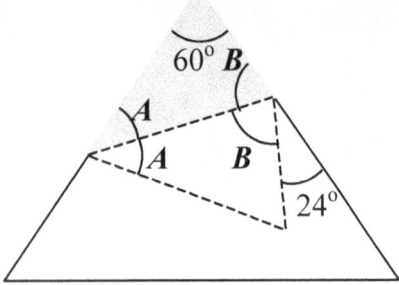

Fig. 2.3

And we found that $A = 42^\circ$.

So we get: $S = 42^\circ \Rightarrow W = 180^\circ - 84^\circ = 96^\circ$.

And we have: $V + T = 180^\circ$, and $T = 67^\circ$. So we get: $V = 180^\circ - T = 180^\circ - 67^\circ = 113^\circ$.

Now, we have: $U + V + W + 53^\circ = 360^\circ$, and $U + A = 180^\circ$.

So first, we get: $U = 360^\circ - V - W - 53^\circ = 360^\circ - 113^\circ - 96^\circ - 53^\circ = 360^\circ - 262^\circ = 98^\circ$.

Thus, next, we get: $U + A = 180^\circ \Rightarrow A = 180^\circ - U = 180^\circ - 98^\circ = 82^\circ$.

And let's next, move on to the figure with the angle **B**, and unfold it the way below:

Fig. 2.4

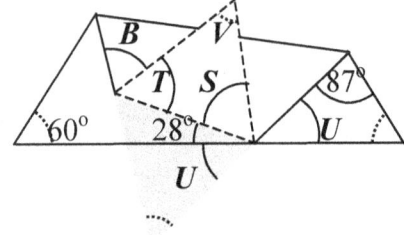

Then first, we can see two identical triangles.
One is in gray, and the other is made of dashed line segments.

So next, we get: $B + 2T = 180^\circ$. So finding T, we can get **B**.
How then, can we find T?

Fully unfolding the figure, we get a regular triangle, where every angle is 60°.
So we can see that $V = 60^\circ$, and thus, can get: $Q + S + 60^\circ = 180^\circ$.
So finding S, we can find Q. How then, can we find S?

We can notice that $S = 28^0 + U$, because unfolding the figure, we get two identical triangles, one is in gray, and the other is made of dashed line segments.
So finding U, we can find S. Well then, how can we find U?

We know that the original triangle is regular. So we get:
$U + 87^0 + 60^0 = 180^0 \Rightarrow U = 180^0 - 87^0 - 60^0 = 180 - 147^0 = 33^0$.

And thus, we can now, get the angle B.

To begin with, we have: $S = 28^0 + U$, so we get: $S = 28^0 + 33^0 = 61^0$.

So next, we get: $Q + S + 60^0 = 180^0 \Rightarrow Q = 180^0 - 61^0 - 60^0 = 180 - 121^0 = 59^0$.

Thus, next, we get: $B + 2T = 180^0 \Rightarrow B = 180^0 - 2T = 180^0 - 118^0 = 62^0$.

And let's next, move on to the figure with the angle C, and unfold it the way below:

Fig. 2.5

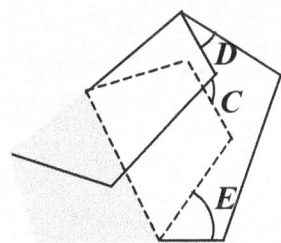

$D = 23^0$, and $E = 45^0$.

Then first, we can see two identical tetragons.
One is in gray, and the other is made of dashed line segments.

Let's consider the tetragon indicated by thick line segments below:

Fig. 2.6

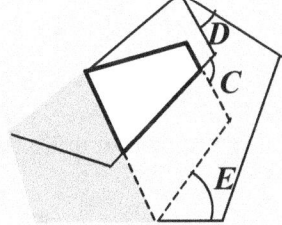

$D = 23^0$, and $E = 45^0$.

And magnifying the tetragon, we can put it the way below:

Fig. 2.7

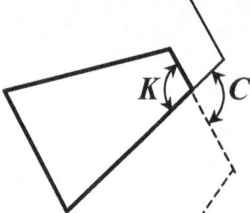

So finding the angle **K**, we get the angle **C**, because **K = C**, since they are vertical angles. How then, can we find **K**?

Fig. 2.8

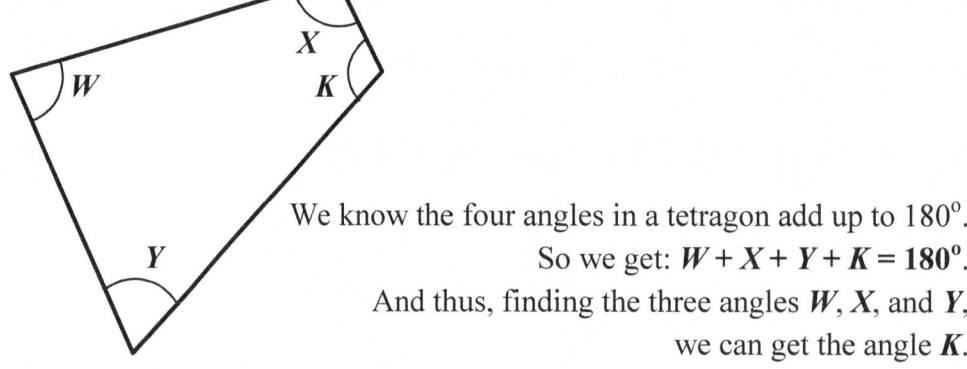

We know the four angles in a tetragon add up to 180°.
So we get: $W + X + Y + K = 180^\circ$.
And thus, finding the three angles **W**, **X**, and **Y**,
we can get the angle **K**.

How then, can we find the three angles?

Let's first, get back to the figure where we can see the two *identical* tetragons.

Fig. 2.9

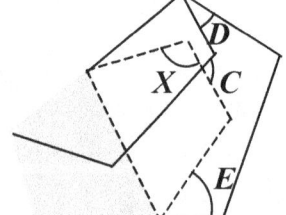

$D = 23^\circ$, and $E = 45^\circ$.

Then, beginning with the angle **X**, what then, can we say about the angle **X**?

The angle **X** is the same as an angle in the tetragon in gray. What angle then, is it?

Indicating two angles by *U* and *V* the way below, we can say that $X = U + V$.

Fig. 2.A

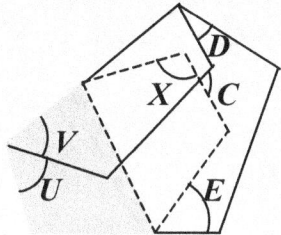

$D = 23^{\circ}$, and $E = 45^{\circ}$.

It's because the gray tetragon is the same as the tetragon made of dashed line segments. How then, can we find the two angles *U* and *V*?

We have already found *U*, which is 49°, which is the solution to the problem 1. What then, about the angle *V*?

We can set: $U + 2V = 180^{\circ}$. How come?

Unfolding the figure above, we can unfold the way to the solution.

Fig. 2.B

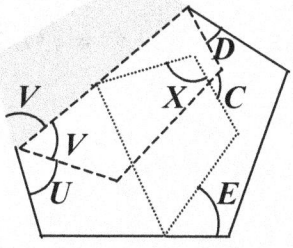

$D = 23^{\circ}$, and $E = 45^{\circ}$.

The tetragon in gray is identical to the tetragon made of dashed line segments.

So we get: $U + 2V = 180^{\circ} \Rightarrow V = (180^{\circ} - U)/2 = (180^{\circ} - 49^{\circ})/2 = 131^{\circ}/2 = 65.5^{\circ}$.

And we know: $X = U + V$. So we get: $X = 49^{\circ} + 65.5^{\circ} = 114.5^{\circ}$.

And let's next, move on to the angle W.

Fig. 2.C 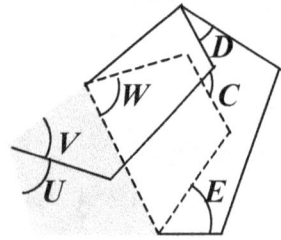 $D = 23^\circ$, and $E = 45^\circ$.

Then again, we can get the angle W using the fact that the gray tetragon is the same as the tetragon made of dashed line segments. How can we use the fact though?

Using the fact, we can name some angles the way below:

Fig. 2.D 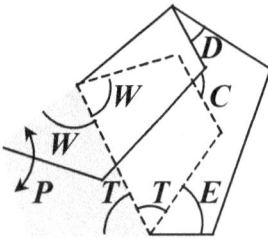 $D = 23^\circ$, and $E = 45^\circ$.

And we know that the sum of all the four angles in a tetragon is 360°, each of the five angles in a regular pentagon is 108°, and of course, the angle $P = U + V = 49^\circ + 65.5^\circ$.

So we get: $W + 114.5^\circ + 108^\circ + T = 360^\circ \Rightarrow W = 360^\circ - T - 222.5^\circ = 137.5^\circ - T$.

So finding T, we can get W. How then, can we find the angle T?

We have: $T + T + E = 180^\circ$, where $E = 45^\circ$.

So we get: $2T + 45^\circ = 180^\circ \Rightarrow T = (180^\circ - 45^\circ)/2 = 135^\circ/2 = 67.5^\circ$.

And thus, we get: $W = 237.5^\circ - T = 137.5^\circ - 67.5^\circ = 70^\circ$.

And let's next, move on to the angle **Y**.

Fig. 2.E

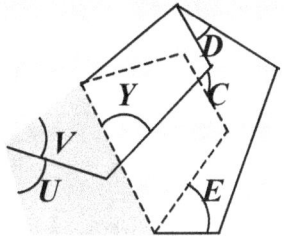

$D = 23^{\circ}$, and $E = 45^{\circ}$.

Magnifying the area where the angle **Y** is, we can get:

Fig. 2.F

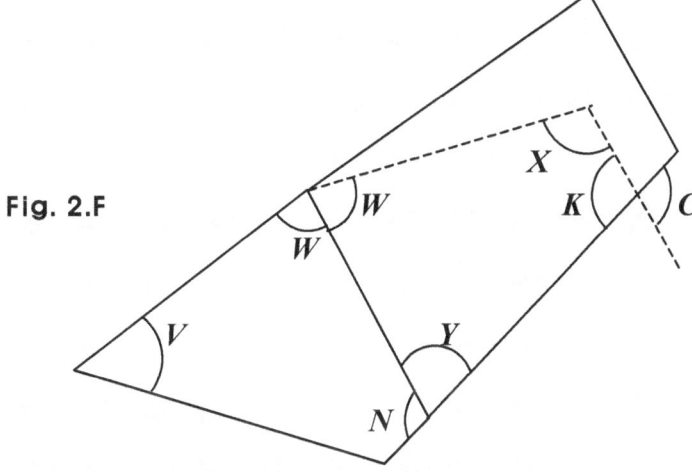

Then, we can see that $N + Y = 180^{\circ}$, and also, that $V + W + N + 108^{\circ} = 360^{\circ}$.
That is, we have: $Y = 180^{\circ} - N$, and $N = 360^{\circ} - 108^{\circ} - W - V = 252^{\circ} - W - V$.
Where is 108° from though?

Every angle in a regular pentagon is 108°, which is in this case next to the angle **N**.
And we know: $V = 65.5^{\circ}$, and $W = 70^{\circ}$, so we get: $N = 252^{\circ} - 65.5^{\circ} - 70^{\circ} = 116.5^{\circ}$.

And thus, we get: $Y = 180^{\circ} - N = 180^{\circ} - 116.5^{\circ} = 63.5^{\circ}$.

Now, we have: $W + X + Y + K = 360^{\circ}$, and also, $K = C$ since **K** and **C** are vertical angles.

So we get: $C = 360^{\circ} - (W + X + Y)$. And we have: $W = 70^{\circ}$, $X = 114.5^{\circ}$, and $Y = 63.5^{\circ}$.

Thus, we get: $C = 360^{\circ} - (70^{\circ} + 114.5^{\circ} + 63.5^{\circ}) = 360^{\circ} - 248^{\circ} = 118^{\circ}$.

Examples 3

0. The sum of all the angles in a triangle is 180°. What then, about the sum of all the angles in each of the polygons as follows: tetragon, pentagon, hexagon, and a polygon with 151 sides?

1. In a polygon, an external (exterior) angle is the angle that is supplement to an internal angle in the polygon, and is adjacent to the internal angle. And if two angles X and Y are supplement to each other, we get: $X + Y = 180^{\circ}$. So for instance, the angle D below is an external angle in the triangle below.

Find the sum of all the external angles for each of the polygons as follows: triangle, tetragon, pentagon, and a polygon with n sides.

Suggestions or Solutions
To the Problem 0

The sum of all the angles in a triangle is 180°. What then, about the sum of all the angles in each of the polygons as follows: tetragon, pentagon, hexagon, and a polygon with 151 sides?

Beginning with a tetragon, we can partition it into triangles either of the ways below:

Fig. 0.0

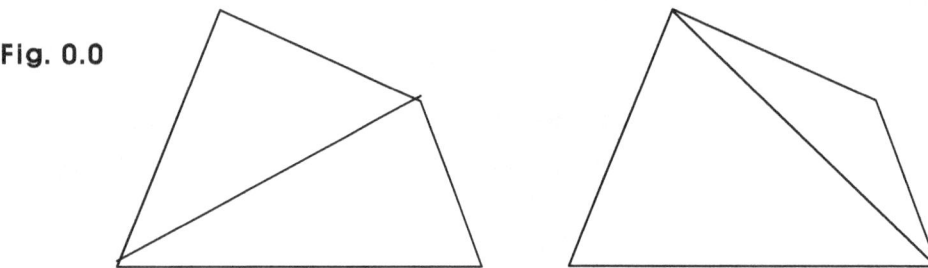

That is, putting together two triangles, we can make a tetragon. Each triangle has three angles, the sum of which is 180°. What then, is the sum of all the angles in a tetragon?

It is the sum of all the angles in two triangles. So the sum is: **2·180° = 360°**.

Next, moving on to a pentagon, we can partition it into triangles many ways:

Fig. 0.1

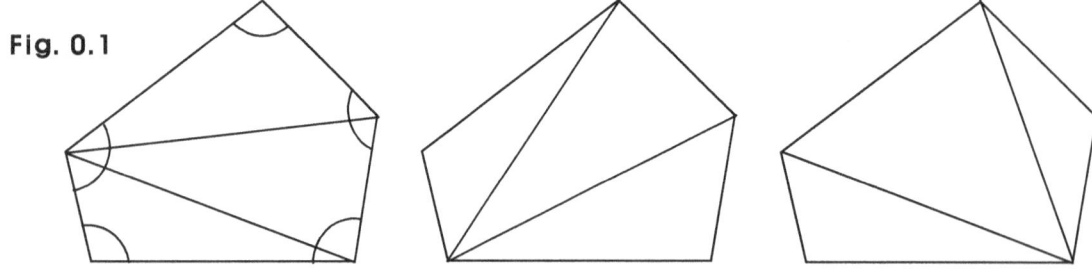

That is, putting together three triangles, we can make a pentagon. Each triangle has three angles, the sum of which is 180°. What then, is the sum of all the angles in a pentagon?

It is the sum of all the angles in three triangles. So the sum of all the angles in a pentagon is: $3 \cdot 180^{\circ} = 540^{\circ}$.

Next, moving on to a hexagon, we can partition it into triangles. That is, putting together triangles, we can make a hexagon. How many triangles then, have to be put together?

Fig. 0.2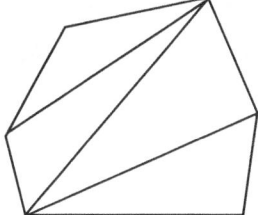

So making a hexagon, we need put together at least four triangles, in each of which, the sum of the thee angles is 180°.

And thus, the sum of all the angles in a hexagon is: $4 \cdot 180^{\circ} = 720^{\circ}$.

What then, about the polygon with 151 sides?

We know:

A tetragon has four sides, and can be made of two triangles.

A pentagon has five sides, and can be made of three triangles.

A hexagon has six sides, and can be made of four triangles.

How many triangles than, do we need to put together to make a heptagon, which is a polygon with 7 sides?

Making a heptagon, we need put together at least five triangles. So the sum of all the angles in a heptagon is: $5 \cdot 180^{\circ}$.

And we can say that making an octagon, which is a polygon with 8 sides, we need put together at least six triangles. So the sum of all the angles in an octagon is: $6 \cdot 180^\circ$.

What then, about the polygon with n sides?

Making a polygon with n sides, we need put together at least $(n-2)$ triangles.

So the sum of all the angles in a polygon with n sides is: $(n-2) \cdot 180^\circ$.

And thus, the sum of all the angles in a polygon with 151 sides is: $149 \cdot 180^\circ$.

Suggestions or Solutions
To the Problem 1

In a polygon, an external (exterior) angle is the angle that is supplement to an internal angle in the polygon, and is adjacent to the internal angle. And if two angles X and Y are supplement to each other, we get: $X + Y = 180^\circ$. So for instance, the angle D below is an external angle in the triangle below.

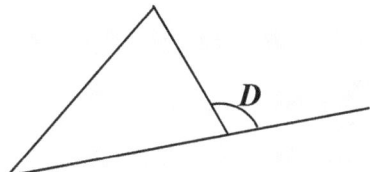

Find the sum of all the external angles for each of the polygons as follows: triangle, tetragon, pentagon, and a polygon with n sides.

We know if two line segments are in a line, the angle between the two is 180°.

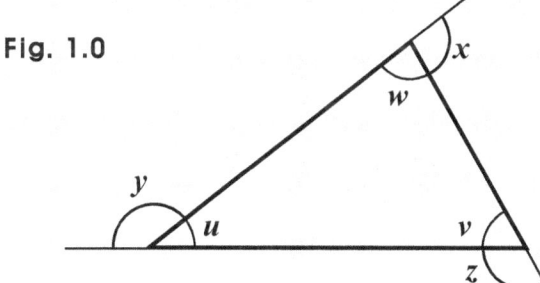

Fig. 1.0

And we know: $x + y + z = 180^\circ$.

So the sum is: $3 \cdot 180^\circ - 180^\circ = 360^\circ$.

And we know that the corresponding angles are the same, so are the alternate angles. So assuming in the figure below, p and q are parallel to each other, we get: $x = s$, and $y = t$.

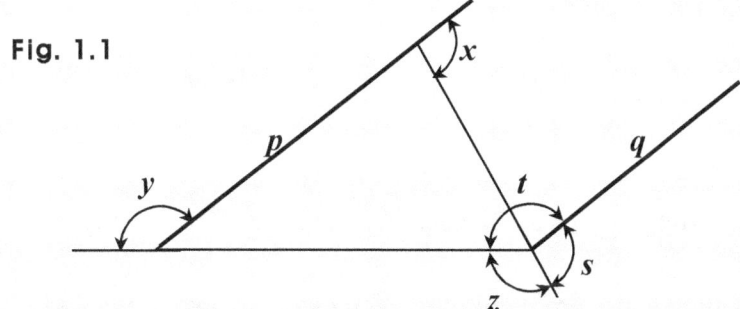

Fig. 1.1

That's because x and s are corresponding angles, and so are y and t.
And thus, the sum is: $x + y + z = s + t + z = 360^\circ$.

Next, moving on to a tetragon, we can partition it into four triangles the way below:

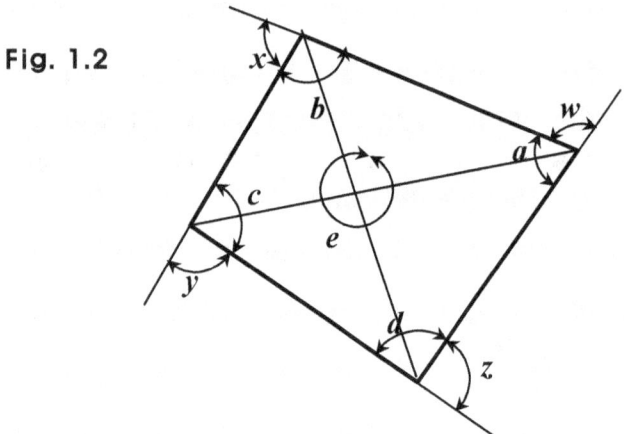

Fig. 1.2

Then first, we can say that:

$e = 360^\circ$, and $a + b + c + d + e = 4 \cdot 180^\circ$, because the three angles in a triangle add up to 180°, and the tetragon is made of four triangles.

So we get: $a + b + c + d = 2 \cdot 180^\circ$.

And next, we can say that: $(a + w) + (b + x) + (c + y) + (d + z) = 4 \cdot 180^\circ$.

In other words, we have: $(a + b + c + d) + (w + x + y + z) = 4 \cdot 180^\circ$.

And we know: $a + b + c + d = 2 \cdot 180^\circ$.

Thus, we get: $w + x + y + z = 2 \cdot 180^\circ = 360^\circ$.

So in the case of a tetragon, too, the sum of all the external angles is 360°.

And we know that the corresponding angles are the same, so are the alternate angles.

So assuming in the figure below, p and q are parallel to each other, and so are m and n, we get: $x = s$, $y = t$, and $z = u$.

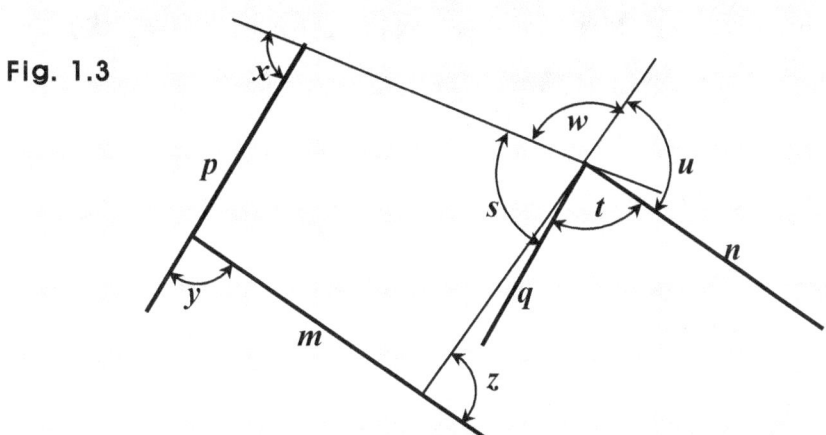

Fig. 1.3

That's because x and s are corresponding angles, and so are y and t, and the same is true for z and u, too. And thus, the sum is: $w + x + y + z = w + s + t + u = 360°$.

Next, moving on to a pentagon, we can partition it into five triangles the way below:

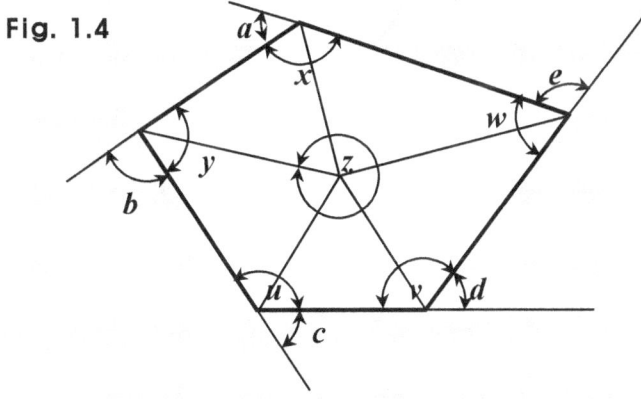

Fig. 1.4

Then first, we can say that:

$z = 360°$, and $x + y + z + u + v + w = 5·180°$, because the three angles in a triangle add up to $180°$, and the pentagon is made of five triangles.

So we get: $x + y + u + v + w = 3·180°$.

And next, we can say that: $(a + x) + (b + y) + (c + u) + (d + v) + (e + w) = 5·180°$.

In other words, we have: $(a + b + c + d + e) + (x + y + u + v + w) = 5·180°$.

And we know: $x + y + u + v + w = 3 \cdot 180^{\circ}$. Thus, we get: $a + b + c + d + e = 2 \cdot 180^{\circ}$.

So in the case of a pentagon, too, the sum of all the external angles is 360°.

And we know that the corresponding angles are the same, so are the alternate angles.

So assuming in the figure below, p and s are parallel, q and m are parallel, and r and n are parallel, we get: $a = x$, $b = y$, $c = z$, and $d = u$.

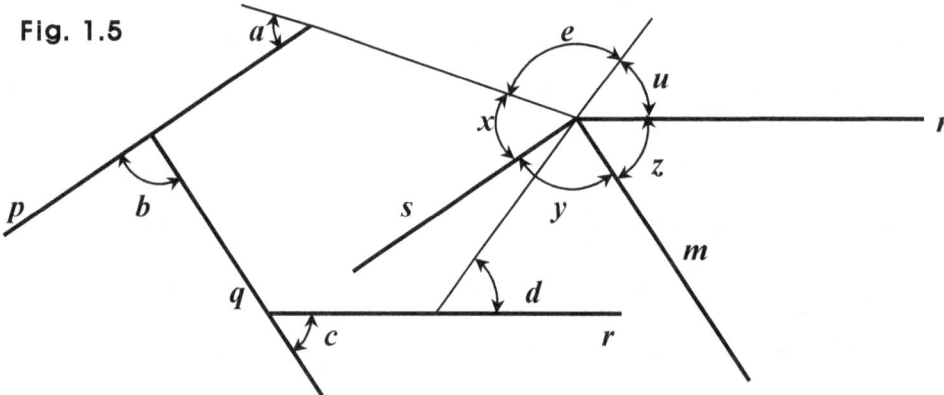

It's because a and x are corresponding angles, and the same is true, too, for b and y, for c and z, and for d and u.

And thus, the sum is: $a + b + c + d + e = x + y + z + u + e = 360^{\circ}$.

Suppose next, seven line segments meet each other at one point the way below:

Fig. 1.6

Then, we can say that the sum of all the angles between the line segments is 360°.

Let's next, put line segments along a circle the way below so that each line segment is tangent to the circle, and is parallel to one of the seven line segments above.

Fig. 1.7

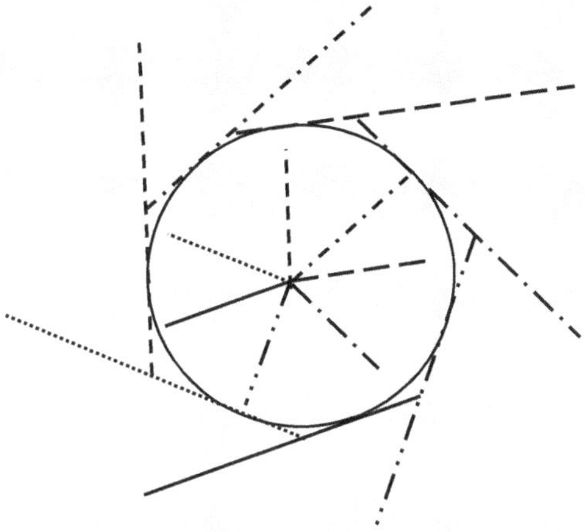

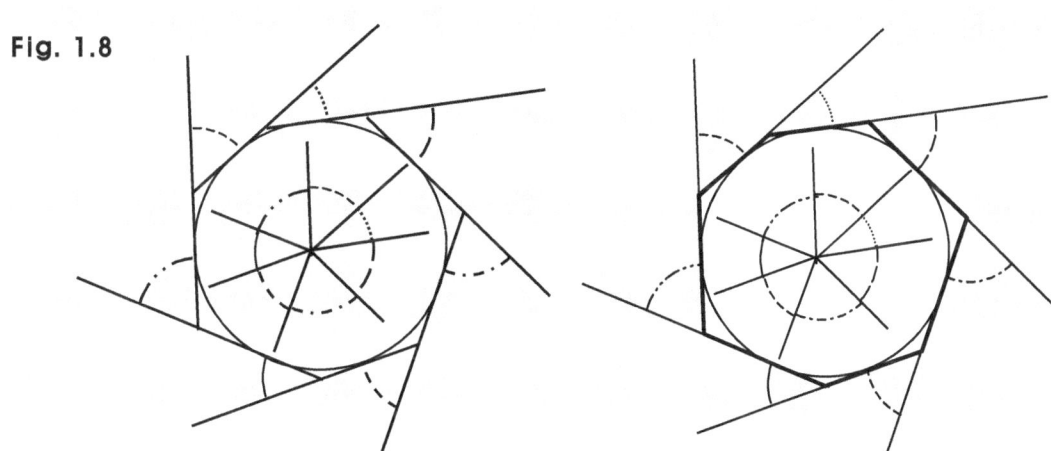

Then, we can indicate pairs of corresponding angles the way below.

Fig. 1.8

Then, we can see a heptagon, and can say that the sum of all the external angles of a heptagon is $360°$.

And thus, we can conclude that taking the sum of all the external angles in every polygon, we get $360°$. That is to say that in a polygon with **n** sides, all the external angles add up to $360°$.

Examples 4

0. Assuming $q = 2p$, and p and q are parallel to each other, find all the triangle that have the same area.

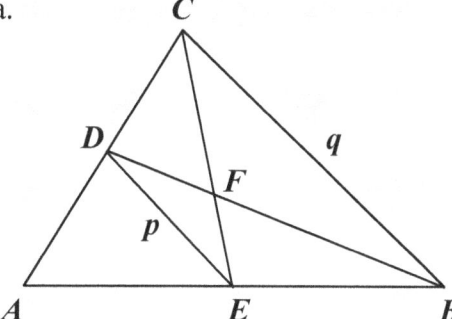

1. Find the conditions under which the two triangles below are identical.

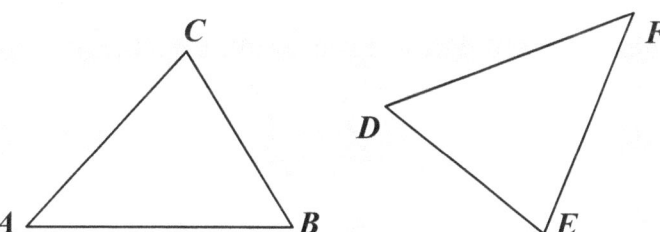

2. Assuming the two triangles below are regular and placed on a line, and meet at one point on the line, find the angle A.

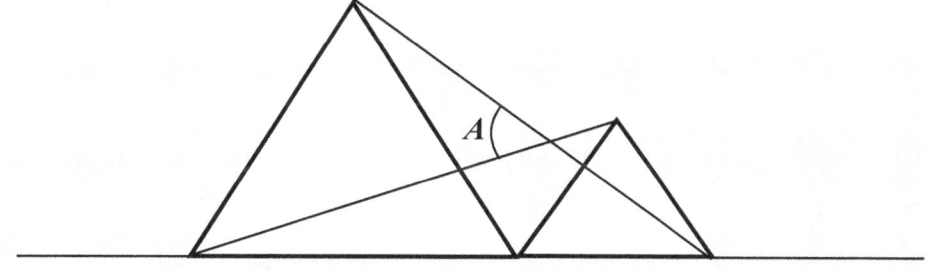

Suggestions or Solutions
To the Problem 0

Assuming in the figure below, $q = 2p$, and p and q are parallel to each other, find all the triangles that have the same area.

Fig. 0.0

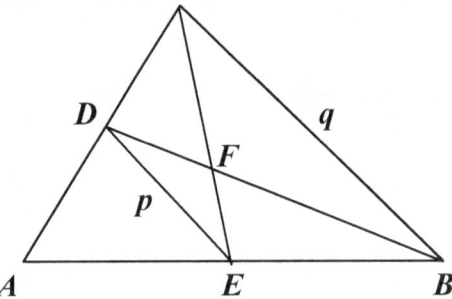

To begin with, what do we mean by the fact that p is parallel to q?

It means that two triangles **AED** and **ABC** are similar to each other.
And if a triangle is similar to another triangle, both have the same set of three angles.

So in this case, we have: $\angle DAE = \angle CAB$, $\angle ADE = \angle ACB$, and $\angle AED = \angle ABC$.

And next, $q = 2p$ means that every side in the bigger triangle is twice its corresponding side. What do we mean by the corresponding side though?

The two sides facing the same angle are said to be corresponding to each other.
In this case, the side **AD** faces the angle **AED**, the side **AC** faces the angle **ABC**, and the two angles are the same because both angles are corresponding angles, since p is parallel to q. So **AD** and **AC** correspond to each other.

Now, we know that every side in the bigger triangle is twice its corresponding side.

So we get: $AB = 2AE$, and $AC = 2AD$ as well as $q = 2p$, which means: $BC = 2ED$.

That is, the ratio between each pair of corresponding sides is the same.

More specifically, the ratio of every side in $\triangle ABC$ to its corresponding side in $\triangle AED$ is the same, and is 2 in this case.

In other words, every side in $\triangle ABC$ is twice its corresponding side in $\triangle AED$.
And we can put it this way, too:

The ratio of every side in $\triangle AED$ to its corresponding side in $\triangle ABC$ is the same, and is 1/2.
That is to say that every side in $\triangle AED$ is half its corresponding side in $\triangle ABC$.
So what does the fact above have to do with this problem?

We know that each side in $\triangle AED$ is half its corresponding side in $\triangle ABC$.
So we can say that D is the midpoint in AC, and E is the midpoint in AB.
That is to say that the area of $\triangle AED$ is the same as $\triangle DEC$. How come?

That's because $AD = DC$, and both triangles share the same height. What then, is the height?

The height can be the line segment that connects the point E and a point in AC, and is perpendicular to AC.

Fig. 0.1

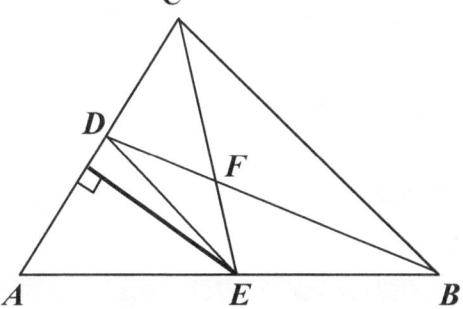

And also, we can say that the area of $\triangle ABD$ is the same as $\triangle DBC$, because $AD = DC$, and both triangles share the same height. What then, is the height?

The height can be the line segment that connects the point **B** and a point in **AC**, and is perpendicular to **AC**.

Fig. 0.2

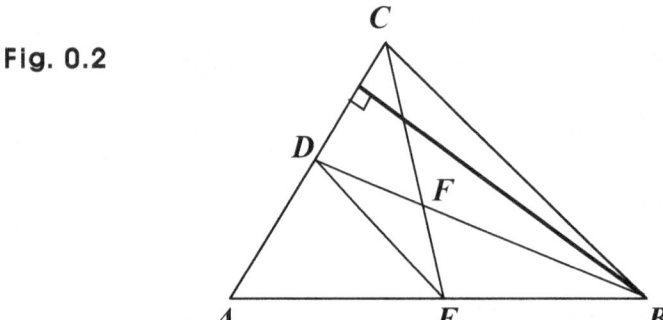

And also, we know that **E** is the midpoint in **AB**.

So we can say that the area of **ΔAEC** is the same as **ΔBEC**, because **AE = EB**, and both triangles share the same height. What then, is the height?

The height can be the line segment that connects the point **C** and a point in **AB**, and is perpendicular to **AB**.

Fig. 0.3

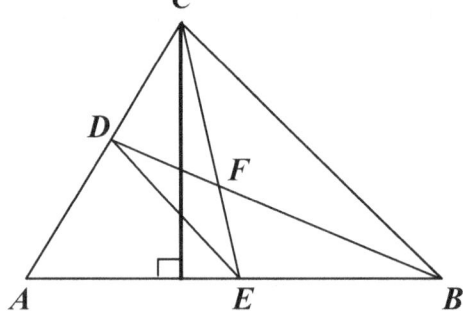

And we can say that the area of **ΔABC** is four times the area of **ΔAED**. How come?

That's because every side in **ΔABC** is twice its corresponding side in **ΔAED**.
More specifically, we can notice that the area of **ΔABC** is twice the area of **ΔAEC**.
And we know that the area of **ΔAEC** is twice the area of **ΔAED**.
So we can say that the area of **ΔABC** is four times the area of **ΔAED**.

Suggestions or Solutions
To the Problem 1

Find the conditions under which the two triangles below are identical.

 **Fig. 1.0**

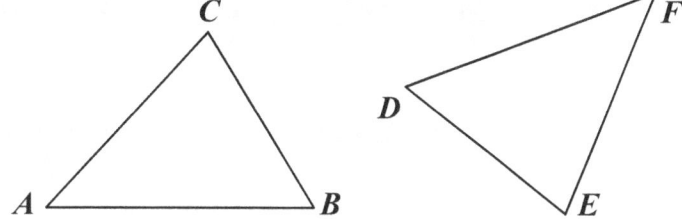

To begin with, if two triangles have the same set of three sides, the two are identical.

So if $AB = EF$, $BC = DE$, and $CA = DF$, we can say that $\triangle ABC$ is identical to $\triangle EFD$.

Fig. 1.1

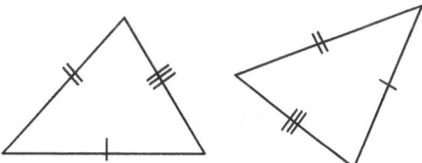

Next, if an angle in one triangle is the same as an angle in the other, and the two sides making the angle in one triangle match the two sides making the same angle in the other triangle, the two triangles are identical.

So in either of the three cases below, we can say that $\triangle ABC$ is identical to $\triangle EFD$.

If $\angle A = \angle F$, $CA = EF$, and $BA = DF$.
If $\angle B = \angle E$, $CB = DE$, and $AB = FE$.   **Fig. 1.2**
If $\angle C = \angle D$, $AC = FD$, and $BC = ED$.

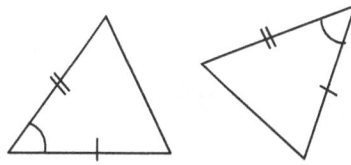

And next, if two angles in one triangle match two angles in the other, and the side between the two angles in one triangle is the same as the side between the two angles in the other triangle, the two triangles are identical.

So in either of the three cases below, we can say that △*ABC* is identical to △*EFD*.

If ∠*A* = ∠*F*, ∠*E* = ∠*B*, and *AB* = *FE*.

If ∠*B* = ∠*E*, ∠*C* = ∠*D*, and *BC* = *ED*.   **Fig. 1.3**

If ∠*C* = ∠*D*, ∠*A* = ∠*F*, and *CA* = *DF*.

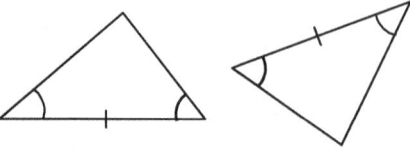

Suggestions or Solutions
To the Problem 2

Assuming the two triangles below are regular and placed on a line, and meet at one point on the line, find the angle A.

Fig. 2.0

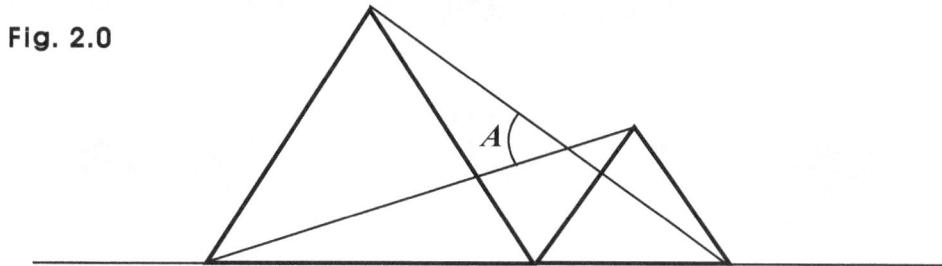

To begin with, in a regular triangle, all the three angles are the same, and so are all the three sides.

Next, adding some labels to the triangles above, we can put them the way blow:

Fig. 2.1

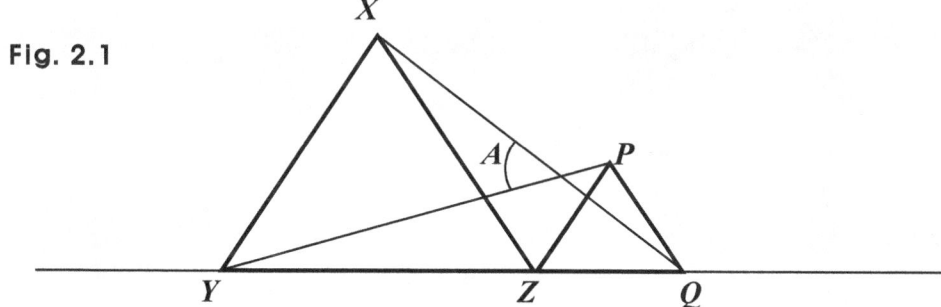

Then, we can notice two triangles identical to each other. One of the two triangles identical is $\triangle YZP$, and the other is $\triangle XZQ$.

And we can put the two triangles $\triangle YZP$ and $\triangle XZQ$ the way blow:

Fig. 2.2

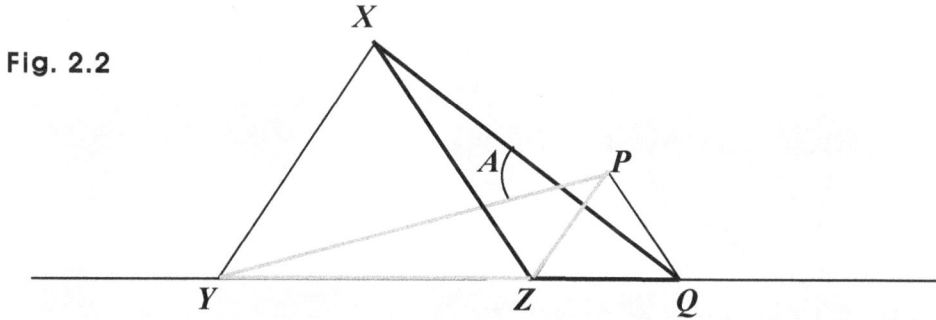

And in fact, turning $\triangle XZQ$ counterclockwise 60° about the point Z, we get $\triangle YZP$.

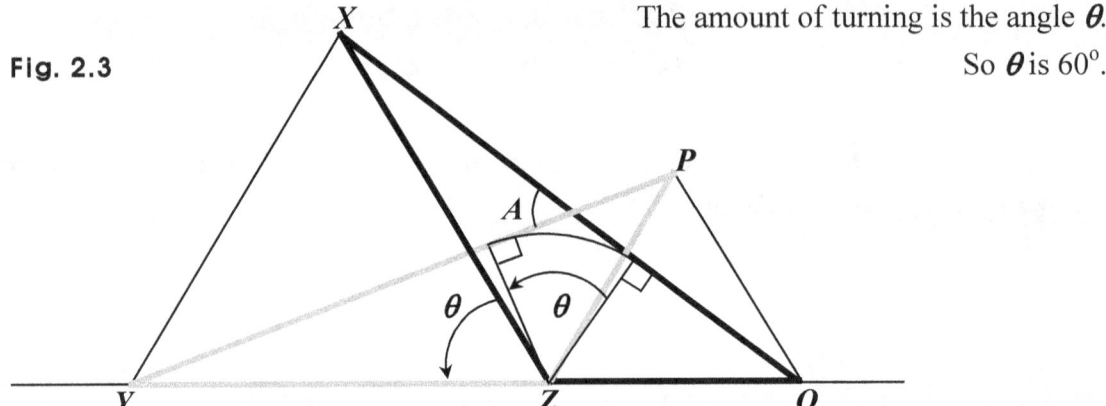

Fig. 2.3

The amount of turning is the angle **θ**.
So **θ** is 60°.

That is to say that turning **XQ** counterclockwise 60° about the point **Z**, we get **YP**.

So we can see that the angle **A** is 60°.
And we can get the same the way below, too:

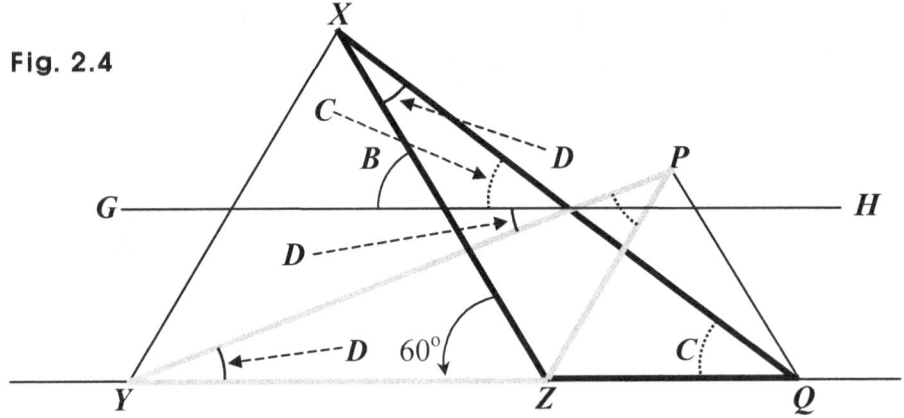

Fig. 2.4

Assuming **GH** is parallel to **YQ**, we can say that **B = 60°**.

And we get: **C + D = B**, because in a triangle, the sum of two internal angles is the same as an external angle supplement to the other internal angle. So in this case, **C + D** is the sum of the two internal angles, and **B** is the external angle.

And we know: **B = 60°**. So we get: **C + D = 60°**.

And also, we can say that **C + D = A**. So we get: **A = 60°**.

Examples 5

0. Assuming **AB** is parallel to **PQ**, **XY** = 5, **VY** = 3, and **YZ** = 12, find **UY**.

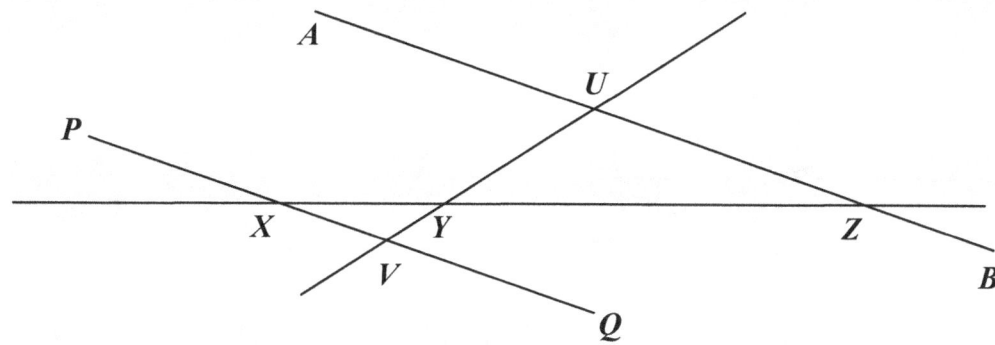

1. Assuming **AB** is parallel to **CD**, **AC** = 2, **AE** = 3, find the ratio between **AB** and **CD**, and the ratio between **BE** and **BD**.

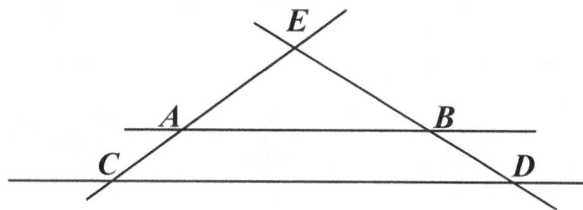

2. Assuming **AE** = 5, **AB** = 3, and **AC** = 2, find the lengths of **BE**, **BD**, and **CD**, and the ratio between the area of the triangle **CDE** and the area of the triangle **BAE**.

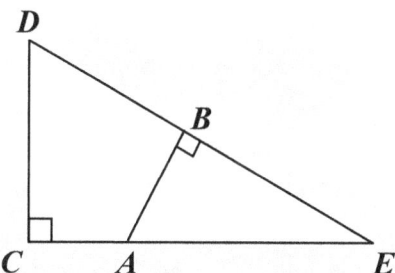

Suggestions or Solutions
To the Problem 0

Assuming in the figure below, *AB* is parallel to *PQ*, *XY* = 5, *VY* = 3, and *YZ* = 12, find *UY*.

Fig. 0.0

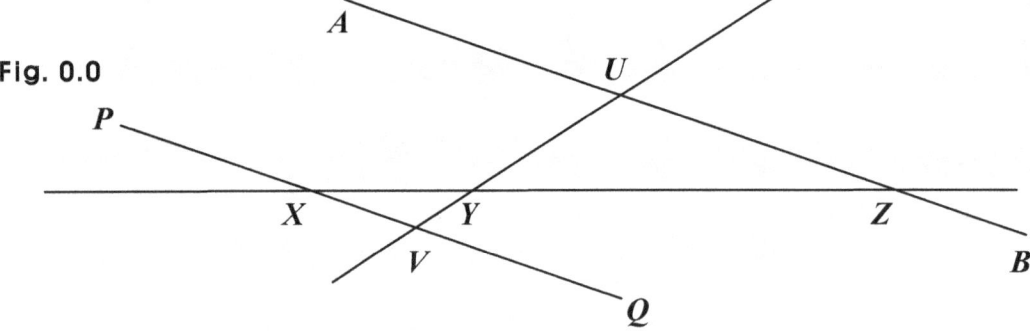

To begin with, we can see two triangles, one is **△*XVY***, and the other is **△*ZYU***.

And we can say that the two are similar triangles. How come?

Since the two lines **AB** and **PQ** are parallel, the two triangles have the same set of three angles, and we can specify all the angles in the two triangles the way below:

Fig. 0.1

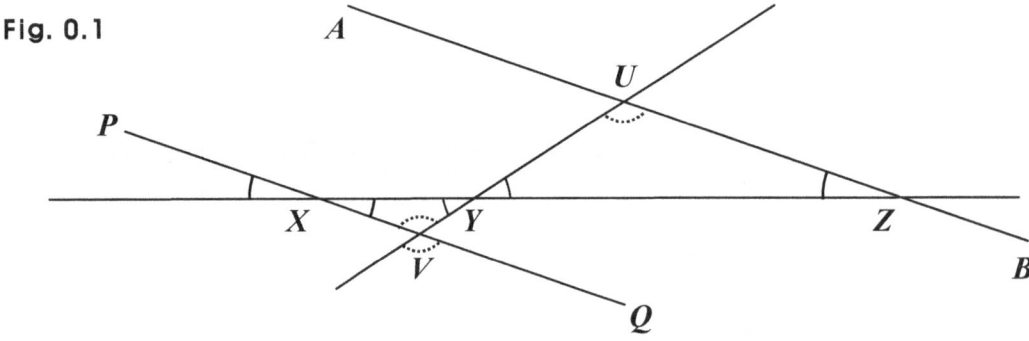

And if two triangles are similar, each pair of corresponding sides keeps the same ratio.

In this case, **XY** corresponds to **ZY**, **VY** corresponds to **UY**, and **VX** corresponds to **UZ**.

So we get: $XY / ZY = VY / UY = VX / UZ$. How then, can we get the length of UY?

We have: $XY = 5$, $VY = 3$, and $YZ = 12$, which is ZY, too, of course.

So we get: $XY / ZY = 5/12 = VY / UY = 3 / UY \Rightarrow 5/12 = 3 / UY \Rightarrow UY = 3 \cdot 12/5 = 36/5$.

Suggestions or Solutions
To the Problem 1

Assuming in the figure below, *AB* is parallel to *CD*, *AC* = 2, *AE* = 3, find the ratio between *AB* and *CD*, and the ratio between *BE* and *BD*.

Fig. 1.0

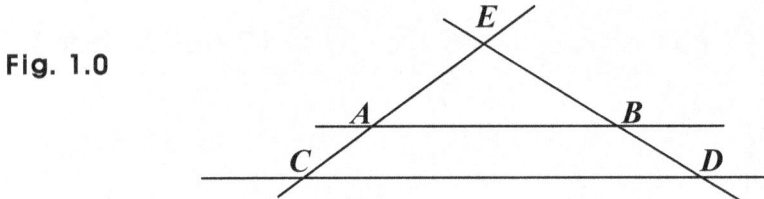

To begin with, we can see two triangles, one is Δ*ABE*, and the other is Δ*CDE*. And we can say that the two are similar triangles. How come?

Since the two lines *AB* and *CD* are parallel, the two triangles have the same set of three angles, and we can specify all the angles in the two triangles the way below:

Fig. 1.1

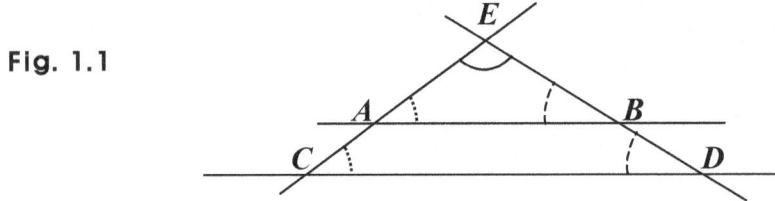

And if two triangles are similar, each pair of corresponding sides keeps the same ratio. In this case, *EA* corresponds to *EC*, *AB* corresponds to *CD*, and *EB* corresponds to *ED*.

So we get: *EA* / *EC* = *AB* / *CD* = *EB* / *ED*.
How then, can we get the ratio between *BE* and *BD*?

To begin with, we have: *AC* = **2** and *AE* = **3**, which is *EA*, too, of course.
So we get: *EA* + *AC* = *EC* = **5**. Thus, we get: *EA* / *EC* = *AB* / *CD* = *EB* / *ED* = **3/5**.

What then, about the ratio of EB to BD, that is, what is EB / BD?

Assuming for instance, $EB = 6$, we get: $EB / ED = 3/5 \Rightarrow ED = EB(5/3) = 10$.

Then, the difference between ED and EB is 4, which is BD.

Then, we get: $EB / BD = 6/4 = 3/2$.

Assuming next, $EB = 18$, we get: $ED = 30$, because $EB / ED = 3/5$.

Then, the difference between ED and EB is 12, which is BD, and we get:

$EB / BD = 18/12 = 3/2$.

How come though, the ratio does not change, that is, EB / BD is still 3/2?

That's because of the distributive law.

For instance, applying the distributive law to subtractions, we can get: $3p - 3q = 3(p - q)$.

Then, we can see that $3p : 3(p - q) = p : (p - q)$, and that $3q : 3(p - q) = q : (p - q)$.

And thus, we can say that the ratio of EB to BD is 3/2, that is, $EB / BD = 3/2$.
What then, about the ratio, EA / AC?

It is the same as EB / BD, which is 3/2. So we get: $EA / AC = 3/2$, too.
That is to say that $EA / AC = EB / BD = 3/2$.

And thus, suppose in the figure below, $AB \text{ // } CD$, that is, AB is parallel to CD, $EC \text{ // } BF$, and $ED \text{ // } AF$.

Fig. 1.2

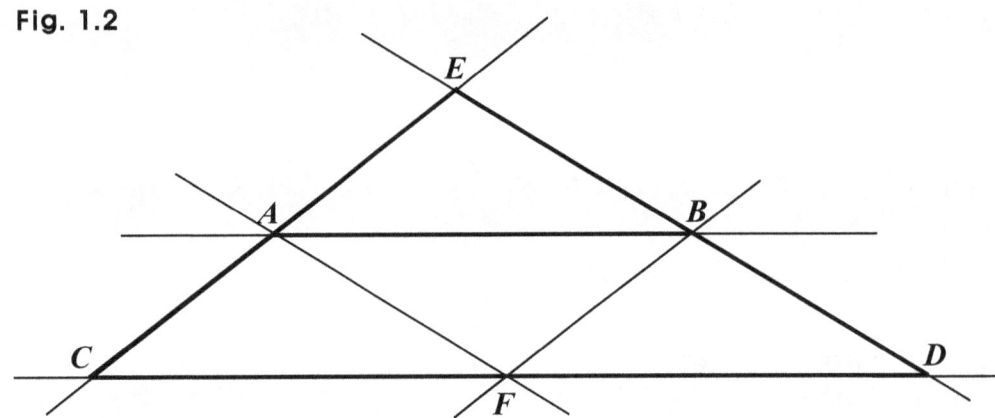

Then, we get:

EA / EC = AB / CD = EB / ED.

EA / AC = CF / FD = EB / BD.

EC / AC = CD / FD = ED / BD.

Suggestions or Solutions
To the Problem 2

Assuming in the figure below, *AE* = 5, *AB* = 3, and *AC* = 2, find the lengths of *BE*, *BD*, and *CD*, and the ratio between the area of △*CDE* and the area of △*BAE*.

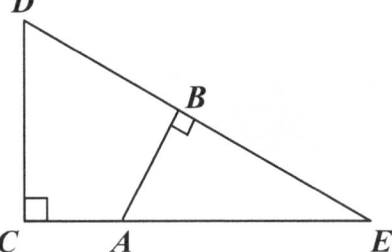

Fig. 2.0

To begin with, we can see two triangles similar, one is △*CDE*, and the other is △*BAE*.

So we can put the figure above more specifically the way below:

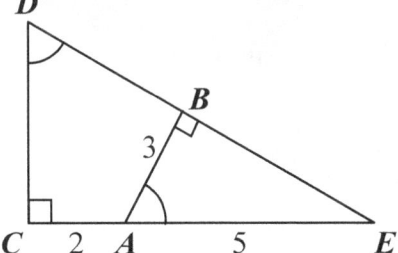

Fig. 2.1

And if two triangles are similar, each pair of corresponding sides keeps the same ratio.

In this case, *AE* corresponds to *DE*, *AB* corresponds to *DC*, and *BE* corresponds to *CE*. So we get: *AE* / *DE* = *AB* / *DC* = *BE* / *CE*.

Then first, how can we get: *BE*, *BD*, and *CD*?

To begin with, △*BAE* is a right triangle, *BE* is one of the two legs in the right triangle. And we know that the other leg is 3, and the hypotenuse is 5.

So assuming $BE = b$, and using the distance formula, we get:

$$5^2 = 3^2 + b^2 \Rightarrow b^2 = 25 - 9 = 16 = 4^2 \Rightarrow b = \pm 4.$$

And we know: $b = BE > 0$. So we get: $BE = 4$.

And let's next, move on to CD.

Then, finding it, we can use: $CE = 7$, and a ratio, which is AB / BE, which is 3/4.

That's because $\triangle CDE$ is similar to $\triangle BAE$. So we get: $DC / CE = AB / BE$. How come?

We have: $AB / DC = BE / CE$.
So multiplying both sides by DC, we get:

$AB = DC \cdot BE / CE$. Next, dividing both sides by BE, we get: $AB / BE = DC / CE$.

Now, we know: $AB / BE = 3/4$, $AB / BE = DC / CE$, and $CE = 7$.

So we get: $DC / CE = DC / 7 = 3/4 \Rightarrow DC = 21/4$.

What then, about BD?

We can put it this way: $BD = DE - BE = DE - 4$. So we get: $DE = BD + 4$.

And we know: $\triangle CDE$ is a right triangle, DE is the hypotenuse, and the two legs are as follows: $DC = 21/4$, and $CE = 7$.

So using the distance formula, we can get DE, and in turn, can get BD.

This time, too, though, we may want to use a ratio AB / AE, which is 3/5.

That's because $\triangle CDE$ is similar to $\triangle BAE$. So we get: $AB / AE = DC / DE$. How come?

We have: $AE / DE = AB / DC$. So multiplying both sides by DC, we get:

$AB = DC \cdot AE / DE$. Next, dividing both sides by AE, we get: $AB / AE = DC / DE$.

Now, we know: $AB / AE = 3/5$, $AB / AE = DC / DE$, and $DC = 21/4$.

So we get: $DC / DE = 3/5 \Rightarrow DE = (21/4)(5/3) = 35/4$.

And we know: $BD = DE - 4$. So we get: $BD = 35/4 - 4 = (35 - 16)/4 = 19/4$.

And let's next, move on to the ratio between the area of $\triangle CDE$ and the area of $\triangle BAE$. How then, can we get the ratio?

We can get the ratio, of course, if we find the areas of the two triangles.
We know however, that the two triangles are similar. So using the ratio between the corresponding sides, we can get the ratio between the two areas. What then, is the ratio?

It is the square of the ratio between the corresponding sides. What then, is the ratio between the corresponding sides?

We have: $AE / DE = AB / DC = BE / CE$. And we know: $BE = 4$, and $CE = 7$.

And we know that BE belongs to $\triangle BAE$, and CE belongs to $\triangle CDE$.

So the ratio of the area of $\triangle BAE$ to the area of $\triangle CDE$ is $4^2/7^2$, which is 16/49.

Examples 6

0.   Assuming **AC = 5**, and **AB = 4**, find the lengths of **BD** and **CD**, and the ratio between the areas of **△BCD**, **△BCA**, and **△CDA**.

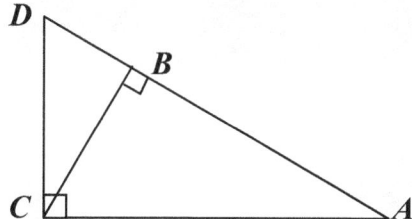

1.   Assuming ∠**ABD** = ∠**ADC**, **BD = 3**, **AD = 8**, and **CD = 5**, find the angle **C** and the length of **AC**.

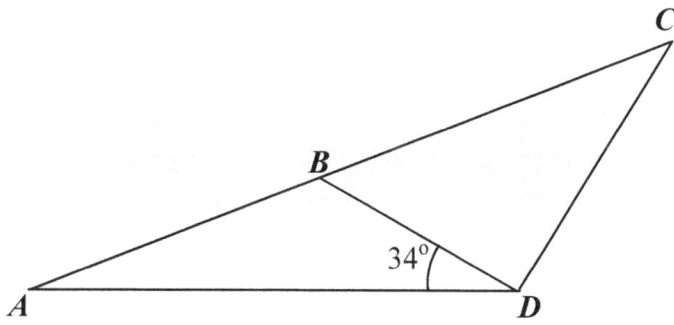

2.   Assuming ∠**ADB** = ∠**ACD**, **AB = 3**, and **AD = 5**, find the angle **C** and the length of **AC**.

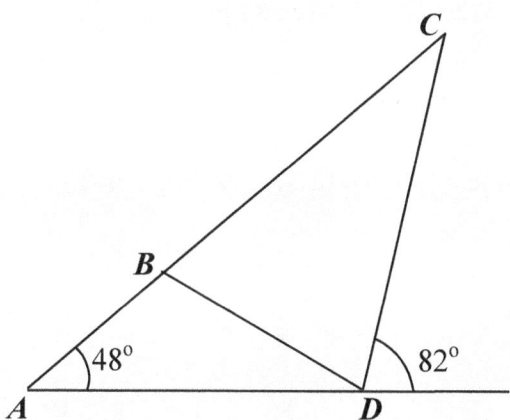

Suggestions or Solutions
To the Problem 0

Assuming in the figure below, $AC = 5$, and $AB = 4$, find the lengths of BD and CD, and the ratio between the areas of $\triangle BCD$, $\triangle BCA$, and $\triangle CDA$.

Fig. 0.0

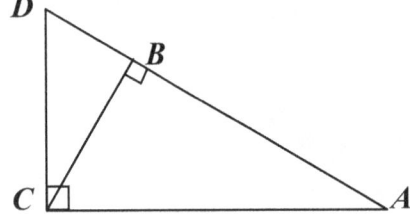

To begin with, we can see that the three triangles given are right triangles.
They are not only so. But they are similar triangles, too.
And if two triangles are similar, each pair of corresponding sides keeps the same ratio.

Looking at $\triangle BCD$ and $\triangle BCA$, we can say that AB corresponds to CB, AC corresponds to CD, and BC corresponds to BD. So we get: $AB / CB = AC / CD = BC / BD$.

How then, can we get BD and CD?

We have: $AC = 5$, and $AB = 4$.

So we get: $AB / CB = AC / CD = BC / BD \Rightarrow 4 / BC = 5 / CD = BC / BD$.

And thus, finding BC, and using $4 / BC = 5 / CD$, we can get CD.

And also, using $4 / BC = BC / BD$, we can get BD.

And of course, after finding CD, and using $5 / CD = BC / BD$, we can get BD, too.
How then, can we find BC?

In $\triangle BCA$, the hypotenuse is AC, which is 5, AB is a leg, and is 4, and BC is the other leg.

So simply using the distance formula, we can get **BC**.

That is, assuming $b = BC$, we get: $5^2 = 4^2 + b^2 \Rightarrow b^2 = 25 - 16 = 9 = 3^2$.

And since $b = BC > 0$, we get: $BC = 3$.

So next, we get: $4 / BC = 5 / CD \Rightarrow CD = (3/4)5 = 15/4$.

And next, we get: $4 / BC = BC / BD \Rightarrow BD = (3/4)3 = 9/4$.

How then, can we find the ratio between the areas of $\triangle BCD$, $\triangle BCA$, and $\triangle CDA$?

We know that for similar triangles, the ratio between the areas is the square of the ratio between the corresponding sides.
And $\triangle BCD$, $\triangle BCA$, and $\triangle CDA$ are similar triangles. How come?

We can put angles in the figure above the way below:

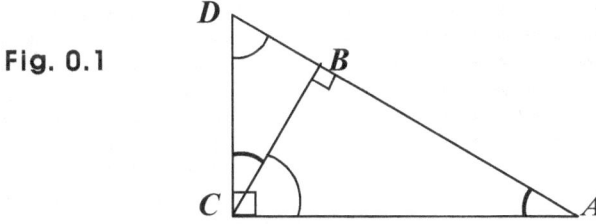

Fig. 0.1

So the three triangles share the same set of three angles.

And assuming **BC** is in $\triangle CDA$, **BA** is in $\triangle BCA$, and **CA** is in $\triangle CDA$, and the three sides **BC**, **BA**, and **CA** are corresponding to each other, we can set $BC : BA : CA = 3 : 4 : 5$.

So assuming p is the area of $\triangle CDA$, q is the area of $\triangle BCA$, and r is the area of $\triangle CDA$, we get: $p : q : r = 3^2 : 4^2 : 5^2$.

Suggestions or Solutions
To the Problem 1

Assuming in the figure below, $\angle ABD = \angle ADC$, $BD = 3$, $AD = 8$, and $CD = 5$, find the angle C and the length of AC.

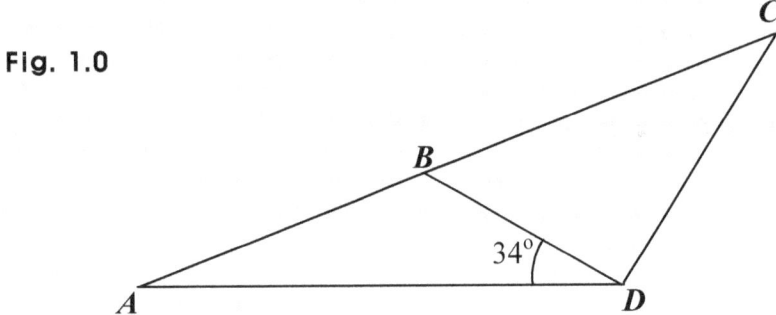

Fig. 1.0

To begin with, we can see two triangles are similar. What are the two though?

Putting in the figure above, the angles given and the sides given, we get:

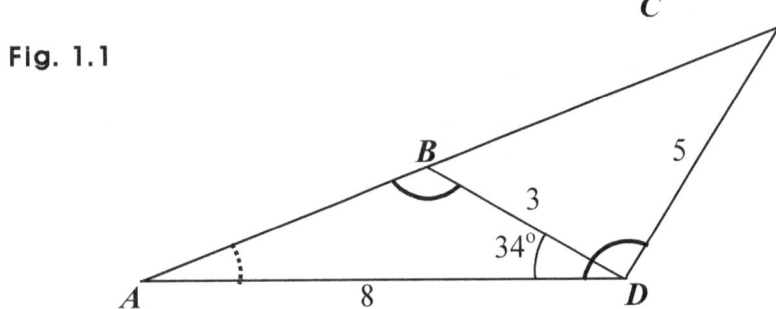

Fig. 1.1

So we can say that $\triangle ABD$ is similar to $\triangle ADC$.
And thus, we can see that the angle C is $34°$. How then, can we get the length of AC?

We know that $\triangle ABD$ is similar to $\triangle ADC$.
So we can say that AD corresponds to AC, AB corresponds to AD, and BD corresponds to DC. Thus, we get: $AD / AC = AB / AD = BD / DC$.
So we get: $8 / AC = AB / 8 = 3/5$.
Thus, we get: $8 / AC = 3/5 \Rightarrow AC = (5/3)8 = 40/3$.

Suggestions or Solutions
To the Problem 2

Assuming $\angle ADB = \angle ACD$, $AB = 3$, **and** $AD = 5$, **find the angle** C **and the length of** AC.

Fig. 2.0

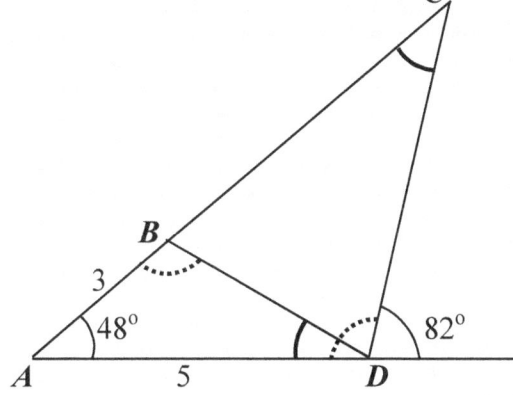

So putting in the figure above, the angles and the sides given, we can see that $\triangle ABD$ is similar to $\triangle ADC$. How then, can we find the angle C?

First, we can say that: $\angle ADC + 82^{\circ} = 180^{\circ}$. So we get: $\angle ADC = 98^{\circ}$.

And we know: $\angle ABD = \angle ADC = 98^{\circ}$, and $\angle ADB = \angle C$.

And also, we know that the sum of the three angles in a triangle is 180°.
So next, in $\triangle ABD$, we can see that $48^{\circ} + \angle ABD + \angle ADB = 48^{\circ} + 98^{\circ} + \angle C = 180^{\circ}$.
Thus, we get: $\angle C = 34^{\circ}$. How then, can we get the length of AC?

We know that $\triangle ABD$ is similar to $\triangle ADC$.
So we can say that AD corresponds to AC, AB corresponds to AD, and BD corresponds to DC. Thus, we get: $AD / AC = AB / AD = BD / DC$.

And we have: $AB = 3$, and $AD = 5$.

So we get: $AD / AC = AB / AD \Rightarrow 5 / AC = 3/5 \Rightarrow AC = (5/3)5 = 25/3$.

Examples 7

0. Assuming $\angle CBD = \angle ACD$, *PQ* is parallel to *AD*, *BD* = 5, and *AD* = 8, find the angle *ABC* and the length of *CD*.

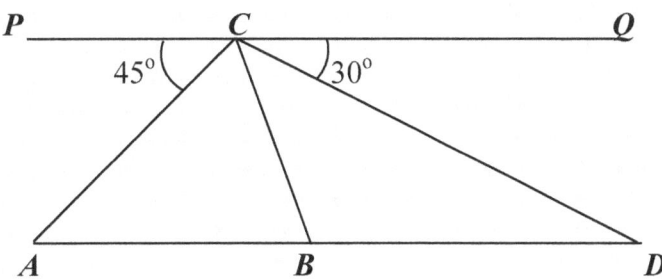

1. Assuming *AB* = 3, *BC* = 8, and *AE* is tangent to the circle, find the length of *AD*.

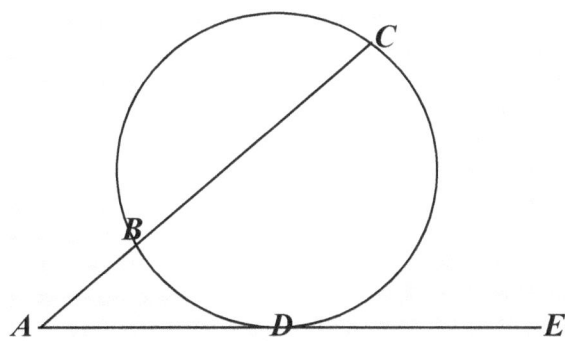

2.   We have a fact that in the figure below, the three angles *X*, *Y*, and *Z* are the same no matter where the three points may be in the circle if they are outside the arc *UV*.

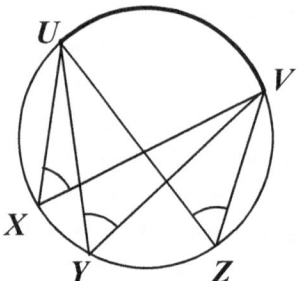

And if the arc *UV* is a half circle, all the angles *X*, *Y*, and *Z* are 90º.

Using the fact above, and assuming in the figure below, *BD* = *DC* = *AD*, *AB* = 3, and the diameter of the circle is 8, find the length of *AD*.

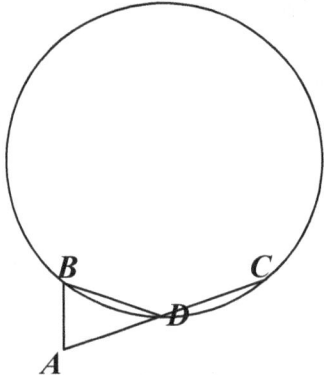

Suggestions or Solutions
To the Problem 0

Assuming $\angle CBD = \angle ACD$, PQ is parallel to AD, $BD = 5$, and $AD = 8$, find the angle ABC and the length of CD.

Fig. 0.0

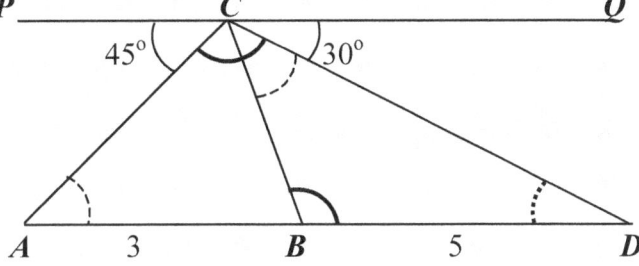

So putting in the figure above, the angles given and the sides given, we can see that $\triangle ACD$ is similar to $\triangle CBD$. How then, can we find the angle ABC?

First, we can say that: $\angle CBD + \angle ABC = 180^{o}$.
So finding $\angle CBD$, we can get $\angle ABC$.
How then, can we find $\angle CBD$?

We know: $\angle ACD = \angle CBD$, and in the figure above, can see: $45^{o} + \angle ACD + 30^{o} = 180^{o}$.

So we get: $\angle ACD = 105^{o}$, and thus, we get: $\angle CBD = 105^{o}$.

So we get: $\angle CBD + \angle ABC = 180^{o} \Rightarrow \angle ABC = 180^{o} - \angle CBD = 180^{o} - 105^{o} = 75^{o}$.

How then, can we find the length of CD?

We know that $\triangle ACD$ is similar to $\triangle CBD$.

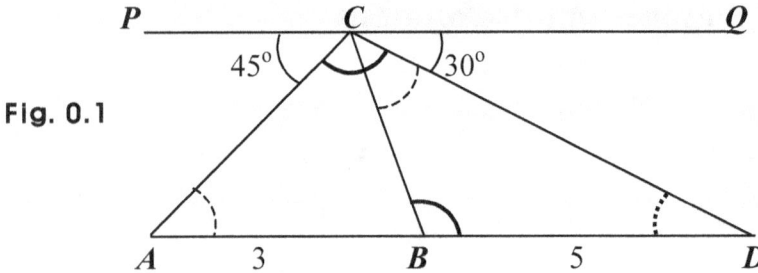

Fig. 0.1

So we can say that *AD* corresponds to *CD*, *AC* corresponds to *CB*, and *CD* corresponds to *BD*. Thus, we get: *AD / CD = AC / CB = CD / BD*.

And we have: *AD = 8*, and *BD = 5*.

So we get: *AD / CD = CD / BD* $\Rightarrow$ *8 / CD = CD / 5*.

And thus, assuming *c = CD*, we get: $c^2 = 40 \Rightarrow c = \pm\sqrt{40} = \pm 2\sqrt{10}$.

And we have: *c = CD > 0*. So we get: $CD = 2\sqrt{10}$.

Suggestions or Solutions
To the Problem 1

Assuming $AB = 3$, $BC = 8$, and AE is tangent to the circle, find the length of AD.

Fig. 1.0

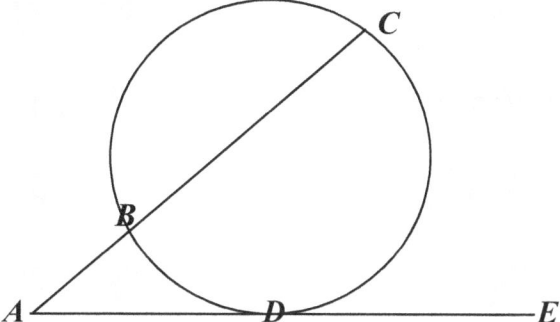

Connecting **B** and **D** and **C** and **D** the way below, we can say $\triangle ABD$ is similar to $\triangle ADC$.

Fig. 1.1

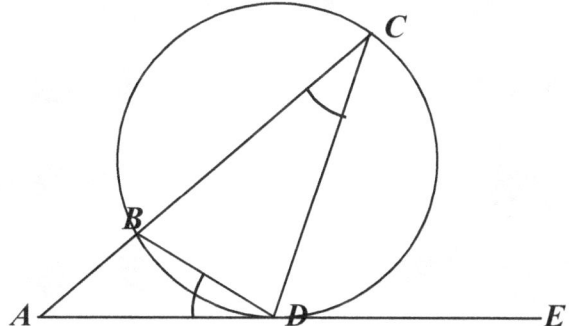

We can say that $\angle A$ belongs to both triangles, and $\angle C$ is the same as $\angle ADB$. How come they are the same though?

Fig. 1.2

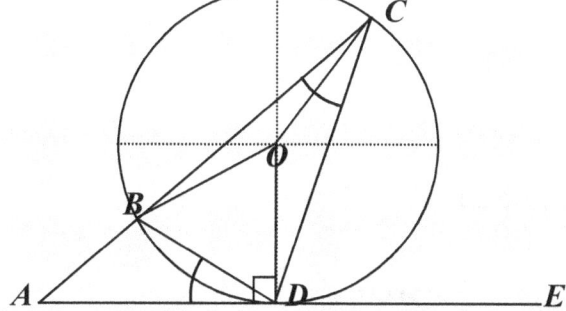

Suppose that the center of the circle is **O**.

Then first, assuming *r* is the radius, we can say that $OB = OC = OD = r$.

So next, we can say that: $\angle OCB = \angle OBC$, $\angle OBD = \angle ODB$, and $\angle OCD = \angle ODC$.

And in a triangle, the three angles add up to 180°. So we can see that:

$\angle OCB + \angle OBC + \angle OBD + \angle ODB + \angle OCD + \angle ODC$
$= 2(\angle OCB + \angle ODB + \angle OCD) = 180^\circ$.

Thus, we get: $\angle OCB + \angle ODB + \angle OCD = 90^\circ$.

And we know that $\angle OCB + \angle OCD = \angle C$. So we get: $\angle C + \angle ODB = 90^\circ$.

And we have: $\angle ADB + \angle ODB = 90^\circ$, too. So we get; $\angle ADB = \angle C$.

Fig. 1.3

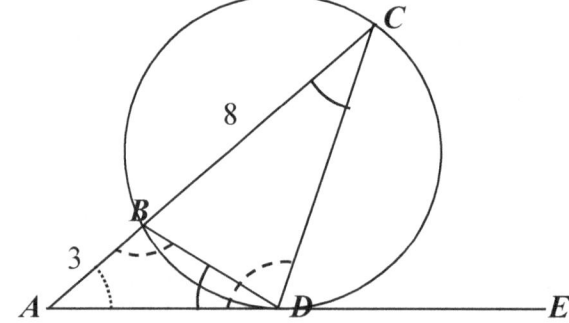

So we can say that $\triangle ABD$ is similar to $\triangle ADC$.

Thus, we get: $AB / AD = AD / AC$.

And we have: $AB = 3$, and $BC = 8$.

So we get: $3 / AD = AD / 11$.

And thus, assuming $a = AD$, we get: $a^2 = 33 \Rightarrow a = \pm\sqrt{33}$.

And we have: $a = AD > 0$. So we get: $AD = \sqrt{33}$.

Suggestions or Solutions
To the Problem 2

We have a fact that in the figure below, the three angles *X*, *Y*, and *Z* are the same no matter where the three points may be in the circle if they are outside the arc *UV*.

Fig. 2.0

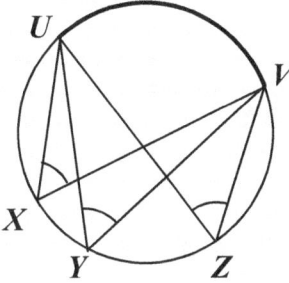

And if the arc *UV* is a half circle, all the angles *X*, *Y*, and *Z* are 90°.

Using the fact above, and assuming in the figure below, *BD* = *DC* = *AD*, *AB* = 3, and the diameter of the circle is 8, find the length of *AD*.

Fig. 2.1

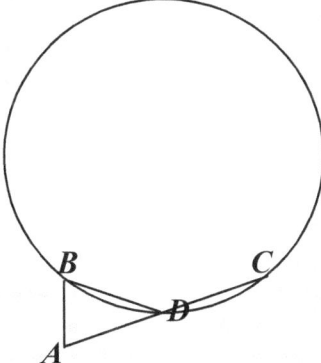

Assuming *FD* passes though the center of the circle, and *E* is the midpoint between *B* and *C*, we can put them the way below:

Fig. 2.2

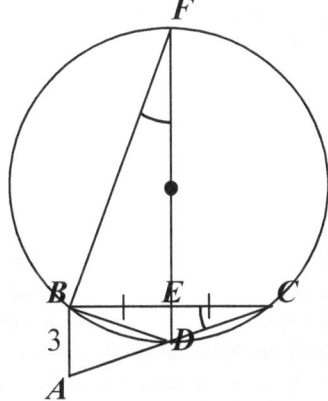

Fig. 2.3

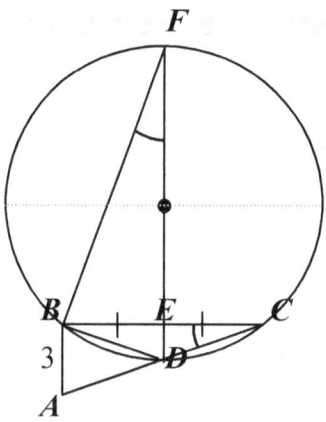

Then first, we can say that **FD** is the diameter.

So using the fact above, we can see that $\angle DAB = 90^{\circ}$.

And also, using the fact above, we can say that $\angle BCD = \angle BFD$, because both cover the same arc **BD**.

Next, we have: **BD = DC = AD**.

So we can see two isosceles triangles, which are **△BCD** and **△ABD**.

And also, we can say that $\angle ABC = 90^{\circ}$. That is, **△ABC** is a right triangle. How come?

To begin with, in **△ABD**, we have: $\angle ABD = \angle DAB$, since the triangle is isosceles.
And in **△BCD**, we have: $\angle DBC = \angle DCB$.

Next, we know in a triangle, the sum of all its angles is 180°.

So taking the sum of all the angles in **△ABC**, we can put it the way below:

$2(\angle ABD + \angle DBC) = 180^{\circ}$.

Thus, we get: $\angle ABD + \angle DBC = 90^{\circ}$.

And we know: $\angle ABC = \angle ABD + \angle DBC$. So we get: $\angle ABC = 90^o$.

And thus, $\triangle ABC$ is a right triangle.

And we know $\triangle DBF$ is a right triangle, too. And also, we know: $\angle F = \angle C$.

So $\triangle ABC$ and $\triangle DBF$ share the same set of three angles, and thus, are similar.

So we can say that $FD \,/\, AC = BD \,/\, AB$.
And we know: $AC = 2AD$, $BD = AD$, $AB = 3$, and $FD = 8$, since FD is the diameter, which is 8.

So assuming $a = AD$, we get:

$$FD \,/\, AC = BD \,/\, AB \Rightarrow 8/(2a) = a/3 \Rightarrow 4/a = a/3 \Rightarrow a^2 = 12 \Rightarrow a = \pm 2\sqrt{3}.$$

And we know: $a = AD > 0$, since a is a length.   So we get: $AD = 2\sqrt{3}$.

www.ingramcontent.com/pod-product-compliance
Lightning Source LLC
Chambersburg PA
CBHW081125170526
45165CB00008B/2547